T0073666

DOES COUNTER-TERRORISM WORK?

'judicious and thoughtful...fair and balanced, critical without being polemical,...relevant and timely. Part of its argument is that history is important, and I am in complete agreement.'

Martha Crenshaw
Senior Fellow Emerita, Stanford University; Professor Emerita, Wesleyan University

'This is a judicious and precisely conceptualized approach to a subject as important and elusive as terrorism itself, but seldom examined so carefully. Policymakers and public alike need to understand why some methods are effective and others are the opposite. This book should become the leading study of this dangerously neglected topic.'

Charles Townshend
Professor Emeritus of International History, Keele University

DOES COUNTER-TERRORISM WORK?

RICHARD ENGLISH

OXFORD
UNIVERSITY PRESS

OXFORD
UNIVERSITY PRESS

Great Clarendon Street, Oxford, OX2 6DP,
United Kingdom

Oxford University Press is a department of the University of Oxford.
It furthers the University's objective of excellence in research, scholarship,
and education by publishing worldwide. Oxford is a registered trade mark of
Oxford University Press in the UK and in certain other countries

© Richard English 2024

The moral rights of the author have been asserted

All rights reserved. No part of this publication may be reproduced, stored in
a retrieval system, or transmitted, in any form or by any means, without the
prior permission in writing of Oxford University Press, or as expressly permitted
by law, by licence or under terms agreed with the appropriate reprographics
rights organization. Enquiries concerning reproduction outside the scope of the
above should be sent to the Rights Department, Oxford University Press, at the
address above

You must not circulate this work in any other form
and you must impose this same condition on any acquirer

Published in the United States of America by Oxford University Press
198 Madison Avenue, New York, NY 10016, United States of America

British Library Cataloguing in Publication Data
Data available

Library of Congress Control Number: 2023942342

ISBN 978–0–19–284334–0

DOI: 10.1093/oso/9780192843340.001.0001

Printed and bound in the UK by
Clays Ltd, Elcograf S.p.A.

Links to third party websites are provided by Oxford in good faith and
for information only. Oxford disclaims any responsibility for the materials
contained in any third party website referenced in this work.

MIX
Paper | Supporting
responsible forestry
FSC® C018072

For CMC, JHE, AFE

Acknowledgements

During the researching and writing of this book, I benefited greatly from discussions of my work following talks that I delivered for the following institutions: the University of California, Santa Barbara; Georgetown University; the University of Edinburgh; the British Academy; the University of Chicago; Yale University; the University of Bristol; the European Consortium for Political Research; the Asser Institute in The Hague; the Council on Foreign Relations; Keene State College; Queen's University Belfast; Boston College; the European University Institute, Florence; the University of Padua; the Conflict Research Society. The advice offered by Andrew Gordon, Luciana O'Flaherty, and Imogene Haslam has proved invaluable.

RE
Belfast
February 2023

Contents

Abbreviations

AQI	Al-Qaida in Iraq
CIA	Central Intelligence Agency
CID	Criminal Investigation Department
CUP	Cambridge University Press
DHS	Department of Homeland Security
DPP	Director of Public Prosecutions
DUP	Democratic Unionist Party
ETA	Euskadi Ta Askatasuna (Euskadi and Freedom, or Basque Homeland and Freedom)
EU	European Union
FBI	Federal Bureau of Investigation
FRU	Force Research Unit
GCHQ	Government Communications Headquarters
GFA	Good Friday Agreement
HMSU	Headquarters Mobile Support Unit
HUMINT	Human Intelligence
ICCT	International Centre for Counter-Terrorism
IDF	Israel Defence Forces
INLA	Irish National Liberation Army
IRA	Irish Republican Army
IRC	Independent Reporting Commission
ISI	Inter-Services Intelligence
ISIS	Islamic State of Iraq and Syria
JTAC	Joint Terrorism Analysis Centre
LHLPC	Linen Hall Library Political Collection
MRF	Military Reaction Force
NIO	Northern Ireland Office
NSA	National Security Agency
OUP	Oxford University Press

PA	Palestinian Authority
PFLP	Popular Front for the Liberation of Palestine
PIJ	Palestinian Islamic Jihad
PIRA	Provisional Irish Republican Army
PLO	Palestine Liberation Organization
PONI	Police Ombudsman for Northern Ireland
PRONI	Public Record Office of Northern Ireland
PSNI	Police Service of Northern Ireland
RPA	Remotely Piloted Aircraft
RUC	Royal Ulster Constabulary
SAS	Special Air Service
SB	Special Branch
SDLP	Social Democratic and Labour Party
SEALs	United States Navy Sea, Air, and Land [SEAL] Teams
SIS	Secret Intelligence Service
TCG	Tasking and Coordination Group
UAV	Unmanned Aerial Vehicle
UDA	Ulster Defence Association
UDR	Ulster Defence Regiment
UFF	Ulster Freedom Fighters
UK	United Kingdom
US	United States
UVF	Ulster Volunteer Force
UWC	Ulster Workers' Council
WMD	Weapons of Mass Destruction

List of Maps

Introduction

I

Terrorism[1] and responses to terrorism remain vital elements in shaping society and politics globally. The central aim of *Does Counter-Terrorism Work?* is to provoke and to help shape debate on an issue which therefore possesses world-changing importance, but which has received less serious and honest discussion than it deserves. From responses to the 1914 terrorist attack which triggered the First World War, to and beyond the 9/11 atrocity, it has been *reactions* to non-state terrorism rather than acts of terrorism themselves which have most decisively changed the world. In that sense, counter-terrorism is an historically even more important phenomenon than terrorist violence. The latter's greatest significance lies not in the damage that is directly done through terrorist attacks, appalling though this is for its victims. Rather, terrorism's far more momentous effects upon politics, society, international relations, economics, and individuals emerge through the reactions that it occasions. In this sense, terrorists have indeed done much to shape the modern world.

Recognition of the mutually shaping relationship between terrorism and counter-terrorism is crucially important. But, if we are fully to comprehend that world-changing relationship, then we also need to understand how far those involved achieve goals which might legitimate their often violent actions. Terrorism is a means of trying to secure important political change. It tends to be justified by its perpetrators as the only or best way of achieving very significant objectives. Hence, we can only claim to understand terrorism when we have ascertained how far such violence does in practice bring about those desired changes.[2] Even more significantly (given states' much larger capacity, endeavour, and influence), counter-terrorism tends to be justified on the ground that it protects people from appalling terrorist violence. Again, therefore, we can only fully understand counter-terrorism if we systematically answer the question that is posed in the title of this book. If countering terrorism is indeed, as UK MI5 Director General Ken McCallum put it in 2020, 'vital work',[3] then understanding the ways in which it is and/or is not effectively pursued is a crucial question—politically, societally, and morally.

Despite this, and in spite of the huge post-9/11 growth in the study of terrorism,[4] the issue of counter-terrorist efficacy has not been satisfactorily explored. Analysis of the effectiveness of terrorism itself has become more prominent in the twenty-first century.[5] But there has not yet emerged a similar understanding of the efficacy of counter-terrorism, and this lack of systematic exploration does us harm if it hampers our ability to respond effectively to terrorist violence.

One central problem has been pithily identified by one of terrorism's greatest scholars, Martha Crenshaw: 'A satisfactory understanding of what constitutes effective counter-terrorism... escapes us.'[6] Straightforward answers are unlikely to suffice, but let's briefly consider a few. Some might suggest that successful counter-terrorism would mean that—for a particular government, police force, or intelligence community over a measurable period of time—the number of terrorist attacks and the number of fatalities and injuries from such attacks could be shown to have diminished. This would certainly be a non-trivial outcome. But such a diminution would at best represent only part of what might be involved in successful counter-terrorism. For example, terrorism famously generates far greater anxiety and reaction than the actual number of its incidents would warrant on coldly proportionate analysis.[7] A diminished rate of attacks, deaths, or injuries might still therefore involve an enduring problem of severe proportions, with one or two easily memorable incidents continuing to undermine public confidence and stability, despite an overall reduction in the number of attacks over a given period. Even a small number of events might still magnify popular fear, depending on their psychological and transgressive qualities. Even more importantly, a reduction of terrorism during a short timeframe (the likeliest basis for the above kind of assessment) might involve state activities which actually lay the groundwork of grievance for future terrorist escalations. In this sense, preventing a wave of attacks might therefore be done in ways that cause later, larger, more frequent terrorist incidents to arise (motivated by a desire for revenge, for example).[8]

Although metrics play an important part in assessing counter-terrorism, they will not therefore answer our question satisfactorily on their own. For example, the reduction in terrorism suggested above would have to be shown to be *caused* by particular counter-terrorist interventions, rather than being generated by autonomous changes in political or economic context or in terrorist organizational capacity, attitude, competence, or strategy. Indeed, the idea of *measuring* the effectiveness of counter-terrorism perhaps involves the wrong phrasing anyway, implying as it does something more open to counting than complex counter-terrorism is likely to prove.

Other existing approaches to the question *Does Counter-Terrorism Work?* have also been limited in ways that this book is intended to remedy.[9] But one broad point to establish at this stage is that the current debate has far too often been a polarized and mutually deaf one. Terrorism understandably provokes strong feelings and reactions, and these have sometimes made more balanced assessments difficult to reach. Many commentators have been relentlessly, unremittingly critical of counter-terrorist efforts. Failure after failure is documented in such analyses, and the negative assessment tends not to be balanced by recognition either of counter-terrorist successes or of the profound difficulties that in practice exist with trying to counter terrorism. In sharp contrast, some analysts have offered somewhat bullish accounts of terrorists being defeated and of great victories by the state. Here, the problem has tended to be myopia about the counter-productive aspects of counter-terrorist endeavour. A persuasive answer to the question *Does Counter-Terrorism Work?* must dispassionately include the achievements as well as the flaws of counter-terrorist efforts.

The importance of such a balanced approach becomes even clearer when we consider some of the implausibly positive self-assessments offered by politicians about counter-terrorist effectiveness. Maintaining strong credibility in counter-terrorist argument is one of the foundation stones for success.[10] But politicians' trumpeting of victorious work has sometimes lacked perspective and authenticity here. President George W. Bush claimed in January 2002 that, 'In four short months, our nation has . . . rallied a great coalition, captured, arrested, and rid the world of thousands of terrorists, destroyed Afghanistan's terrorist training camps. . . . Terrorists who once occupied Afghanistan now occupy cells at Guantanamo Bay. And terrorist leaders who urged followers to sacrifice their lives are running for their own.' Such claims repeatedly were made by the president. In September 2006, he observed: 'Thanks to the hard work of our law enforcement and intelligence professionals, we have broken up terrorist cells in our midst and saved American lives. Five years after 9/11, our enemies have not succeeded in launching another attack on our soil.'[11] As this book will show in detail, however, such claims were extremely simplistic. They represented only part of the story of counter-terrorism's record in the post-9/11 War on Terror.

Similarly celebratory assessments of counter-terrorism can be found among politicians elsewhere. Former Israeli Prime Minister Ehud Olmert: 'We have developed capabilities that no other country in the world has, and I say this with certainty. Not the United States or Russia, and certainly not any other country. The capabilities we developed broke the backbone of the main terrorist efforts after the year 2000'; former Israeli Prime Minister Benjamin Netanyahu, on whether military methods could eliminate terrorism: 'Why not? Certainly.'[12] Again, this book will demonstrate the one-sided exaggerations involved in such assessments.

What is needed as we reflect on counter-terrorist effectiveness is therefore a balanced rather than a tendentious approach. Such analysis must be based on a systematic framework of understanding: a layered definition of what effective counter-terrorism would actually involve in complex reality (something also lacking from existing accounts). Such an approach will involve applying that framework through cross-case, historically informed comparison of lived experience.

This book combines case-study detail with synoptic, wider-angled understanding, the latter involving consideration of family resemblances between the particular contexts that are studied in the three Parts. The study adopts two particularly important aspects of methodological approach in the form of: (first) a systematic, layered framework for understanding what 'working' might mean in relation to a phenomenon as complex as counter-terrorism; and (second) an attitude of long-termism and historical-mindedness. I will briefly set out each of these two elements in turn, before deploying both throughout the book.

II

In considering the effectiveness of counter-terrorism, we need a systematic and comprehensive framework within which to assess what it would actually mean for counter-terrorism to 'work'. Given the messy complexity of human endeavour in such important arenas, such a framework needs to be multi-levelled, its various layers allowing for nuanced understandings of what 'working' might involve. I have suggested that terrorism and counter-terrorism exist in a mutually shaping relationship, and therefore I propose that we adopt the framework of analysis for counter-terrorism that I have used when considering the related question 'Does Terrorism Work?'.[13] Some will find this provocative, since they resist assumptions of echoes between non-state terrorists and their state opponents. But this book will argue that there are indeed four (sometimes reinforcing, sometimes mutually hostile) layers of potential success in counter-terrorism, as there are in terrorism.

Accordingly, I will argue that in counter-terrorism there could be:

(1) strategic[14] victory, with the achievement of a central, primary goal or goals;
(2) partial strategic victory, in which:
    (a) one partially achieved one's central, primary goal(s);
    (b) one achieved or partially achieved one's secondary (rather than central, primary) strategic goal(s);
    (c) one determined the agenda, thereby preventing one's opponent from securing victory;
(3) tactical success, in terms of:
    (a) operational successes;
    (b) the securing of interim concessions;
    (c) the acquisition of publicity;
    (d) the undermining of opponents;
    (e) the gaining or maintaining of control over a population;
    (f) organizational strengthening;
(4) the inherent rewards of struggle as such, independent of central goals.

Within this framework, there is a breadth of things that 'working' could mean in relation to counter-terrorism. *Strategic victory* might involve a state effectively removing the threat of more than trivial terrorism from its territory and people, and thereby getting rid of (almost all) terrorist violence. Perhaps more realistically, it might see a particular, major terrorist organiza-tion becoming effectively neutralized as a danger. Such success could involve the repression or the imploding failure of a significant terrorist group.[15] It might also involve a near-comprehensive political resolution of the conflict from which terrorism had emerged as a violent symptom. This latter possibility reflects the crucial point that, just as terrorism involves political violence, so too counter-terrorism is a firmly *political* activity. To separate counter-terrorism from politics in our analysis would lead to a profound misunderstanding of the subject.

Relatedly, strategic success of the kind set out above could see the state re-establishing its Weberian monopoly over legitimate violence within its territory. This also points to the importance of the use or threat of force within state power, not least in relation to counter-terrorism. Coercion alone will be weakened without widespread popular legitimacy and consent. Here, as elsewhere, serious reflection on counter-terrorism therefore involves questions of political meaning and power.

In addition to strategic victory, state counter-terrorism might involve *partial strategic victory*, in which a state significantly reduced terrorist capacity and lethality, and substantially managed to protect the security of its people. Though not as full a triumph as strategic victory, this second level of success would largely fulfil one of the truly major tasks of state governments, that of order maintenance.[16] Even with some terrorist threat persisting at more than trivial level, a normality-sustaining containment of the danger would represent a partial strategic victory, as would the achievement or partial achievement of the state's secondary goals in counter-terrorism, objectives such as reassuring the public that the state will endure, or persuading people that a particular government deserves strong support, or convincing people that the state (or a particular leader or government within it) deserves one's backing, or convincing a population that the state is not only legitimate but also able to protect them. Here, the endurance of the state (and perhaps especially the continuity of its current boundaries) could embody a significant partial strategic success. But partial strategic success might also be achieved if the state determines the relevant political agenda: if terrorists' narratives are prevented from gaining traction, and if terrorists' own strategic goals are thwarted and their victory therefore prevented.

In addition to strategic and partial strategic victory, counter-terrorism might be judged to have worked also at the *tactical* level. This sometimes reinforces, and at times can work against, strategic successes. Tactically, there can be operational successes (such as the thwarting of a particular attack). There might be interim concessions (a temporary ceasefire, or the release of hostages held by terrorists). There could be the generation of good publicity in relation to the struggle between state and non-state actors, or the undermining of terrorist opponents (in terms of their finances, for example, or their levels of recruitment). Control of a population might emerge (in terms of operational state power across the relevant territory, and the possession of effective control despite the challenge of terrorist organizations). There might also be organizational strengthening (with state resilience being reinforced, for instance, through effective defence against terrorist attack, or with better-funded and more coordinated wings of the state in relation to the fight against terrorism).

Beyond the strategic, the partial strategic, and the tactical, counter-terrorism could also be judged to work if it offered its practitioners *inherent rewards*. People involved in counter-terrorism might gain benefits in terms of career success, reputation, financial gain, or psychological and emotional benefits. Quite reasonably, the literature on terrorism often stresses the normality of those who are involved in terrorist activity.[17] The normality of state actors is less often stressed. But it is at least as important, and has now begun to be more recognized within the literature.[18] The state is multi-faceted and complex, and inherent rewards might be accrued by members of governments, the police, intelligence services, the military, state broadcasters, the prison system, the judiciary, state-run industries or businesses, and any relevant part of the civil service bureaucracy. Inherent rewards might be those involved in a successful career, and in the promotions or financial and other returns that can be associated with such careers. They might be psychological (the rewards of vocationally righteous work against what is considered a pernicious enemy; the satisfactions available through revenge, or excitement, or comradeship, or mutual loyalty, or fame, or high esteem). There can be more going on in counter-terrorism than merely the countering of terrorism, and different kinds of people can have a good counter-terrorist war (whatever the strategic outcome eventually reached). It should also be noted that state efforts to combat, thwart, limit, contain, or defeat non-state terrorism can involve partnership with non-state individuals and groups as well; such relationships can offer further inherent rewards for those involved.

The above framework makes clear that it will rarely be the case that major counter-terrorism campaigns have simply *either* worked *or* failed. As so often, avoiding binary assessments probably makes good historical sense.[19] It might, for example, be the case that many people during a campaign of state counter-terrorism derive attractive rewards (career success, money, celebrity, comradeship, purpose). It might also hold true that the work in which they are involved does secure some notable victories at tactical level (preventing particular terrorist attacks; eroding the capacity of a terrorist organization on which one is focusing attention; establishing better training within, and more coordination between, state organizations which combat terrorism). Notwithstanding all this, at a strategic level the period in question might still be judged to have been partly futile and even counter-productive, with an increase in terrorist attacks and associated fatalities and injuries, with the escalation of certain kinds of organizational threat, and with the under-mining of the state's legitimacy as a credible actor. It is not automatically wrong-headed to consider whether this, for example, is what occurred for the United States and the United Kingdom during the period after the terrorist atrocity of 9/11.

Tactical successes that possess inherent rewards will sometimes make strategic counter-terrorist success more likely; sometimes they will not. As with any human activity, multiple motivations can be involved in counter-terrorism, some of them rendered more explicit than others. Just as multi-causality lies behind counter-terrorist activity, so too any full account of the efficacy of such work will need to acknowledge the positive, negative, and ambiguous qualities of outcomes. If this framework helps us avoid simplistic yes or no answers to the question *Does Counter-Terrorism Work?*, then it also has the advantage of avoiding a situation in which people talk past each other. Those who criticize counter-terrorists for worsening their state's strategic position regarding terrorism, and those who celebrate tactical-operational successes against terrorist adversaries, might shout past each other while yet both being right.

If counter-terrorism is understood to mean efforts by the state to combat, thwart, pre-empt, limit, contain, defeat, or eradicate non-state terrorism,[20] then the above framework offers a systematic way of consistently, coher-ently, and honestly assessing the efficacy of such work. Through case studies and some wider-angled reflection, that is what this book will attempt.

III

In doing so, it will adopt an historically minded approach. This is not because history[21] offers superior insights or understanding compared with other disciplines. Rather, it is necessary both because history offers distinctive approaches towards understanding the concept, causes, and consequences of terrorism,[22] and also because historians have been less salient within debates on terrorism and counter-terrorism than have adherents to numerous other disciplines.[23] It might seem particularly appropriate that responses to terrorism—even more than terrorism itself—should be scrutinized by historians: if 'the major question of history' is indeed 'the transformations of human kind',[24] then the ways in which states have responded to terrorism clearly merit sustained attention by historians, whether in relation to the emergence of the First World War, the process of twentieth-century decolonization, the 9/11 wars, or beyond. Influential and highly relevant figures have pointed out the significance of historical understanding to current-day conflicts and policy, and indeed have stressed the importance of thinking historically if one is to generate effective counter-terrorism.[25] Resonant with such views, this book argues that we need to think about counter-terrorism more historically than has often tended to be the case in practice, and that there are five main aspects of such a mode of thinking that are especially important to consider.

First, we need to assess the efficacy of counter-terrorism with an eye to long pasts (and, by implication, to long futures). Governmental timelines are notoriously short-term. But, just as major terrorist organizations understand their causes and their campaigns in light of long-term dynamics, so too any serious assessment of counter-terrorism must escape the solipsism of the present. This can focus our attention on highly important and policy-relevant questions. How far did the 9/11 atrocity genuinely represent a major fault-line in global politics (as crucial political figures frequently claimed)?[26] Again, is it true that (as is often assumed to be the case) twenty-first-century terrorism represents a more frequent, dangerous, and higher-level threat than that of previous periods?[27] Whether it is a case of assessing the true effects of particular counter-terrorist initiatives, or a matter of understanding how new certain aspects of state counter-terrorism have actually been, it is vital to assess the relationship between change and continuity with an eye to long pasts.

An historically minded approach to understanding the efficacy of counter-terrorism will, secondly, respect the complex particularity of each individual context that is under scrutiny. Historians' attention to context—to very specific settings, places, and times—offers invaluable insights. It is very much counter-terrorisms that we need to understand before we can make compelling claims about counter-terrorism as such. So historical attention to concentric circles of context (individual, small-group, regional, national, transnational) will be significant here.[28] Neither terrorists nor counter-terrorists are homogeneous groups, and nor do they exist beyond explanatory contextual settings of lived experience as it evolves over time. Part of this contextual attention in relation to counter-terrorism will involve clarity about different wings of the state (whether intelligence agencies, police services, the military, government, the civil service, the judiciary, or others) and their respective contextual settings.

Third, an historical approach will involve engagement with a large range of mutually interrogatory sources, including first-hand accounts such as participants' own writings and speeches, contemporary newspapers, archives, interview material, state documents, organizational reports, and other significant primary material. Drawing on such a range of sources does not form the basis for any naive sense of the objective judgement that might arise. No analyst comes to their subject without presuppositions and biases. But listening to so many voices does increase the likelihood of producing detailed, authentic, verifiable assessments that can be as balanced and inclusive as possible; and this is something of very high value in relation to a subject as divisive and controversial as counter-terrorism. It is a frequent lament in the study of terrorism that those analysing it lack adequate data;[29] one of the advantages of an historical approach (both in terms of engaging with long pasts, and also in relation to this point about wide-ranging and primary source material) is that it offers the chance to address that problem. The vast amount of data actually available to us concerning terrorism's long past allows for a fuller knowledge of the phenomenon and of responses to it than would be available through more amnesiac avenues of approach. It is also part of the historian's method to reflect stringently upon the origin and intention behind the production of each source (again, an especially helpful approach in relation to a subject as prone to starkly divergent views, biases, and passions as counter-terrorism).

Fourth, historians tend to be sceptical about overly neat theoretical models of explanation. Procrustean attempts to explain messy human activity in accordance with rigid patterns need to be complemented by consideration of the detailed, historical-contextual circumstances of lived experience. The historical emphasis embodied in this book is therefore intended to offer a constructive counterpoint to other (perhaps more mechanistic) modes of valuable analysis.[30]

Fifth, historians tend to prefer contingency over inevitability when explaining human activity. Deep historical reflection regarding human change over time points away from inevitability, teleology, inexorability, and the predetermined, and instead towards unpredictability, open-endedness, uncertainty, chance, the unexpected, and the contingent.[31] Such insights are reinforced by shrewd economic analyses, such as that which recognizes the existence of 'a radically uncertain world, in which probabilities cannot meaningfully be attached to alternative futures'; 'We can construct narratives and scenarios to describe the ways in which technology and global politics might develop in the next twenty years; but there is no sensible way in which we can refine such dialogue by attaching probabilities to a comprehensive list of contingencies'; 'Good strategies for a radically uncertain world acknowledge that we do not know what the future will hold.'[32] Recent decades have seen a valuable re-emergence within historiography of debates about the contingent and the counter-factual,[33] and historical appreciation of the multiple possibilities that existed in the past offers profound value for our understanding of counter-terrorism.

Clearly, there are aspects of these approaches that are shared by scholars from other disciplines and analysts from other backgrounds. But the combination of these approaches (an assessment of change and continuity in relation to long pasts; a respect for the complex particularity and ultimately the uniqueness of each contextual setting; an engagement with a vast range of mutually interrogatory sources, including first-hand materials; a hesitancy about the explanatory power of rigidly abstract theorizing; and a sympathy for the contingent rather than the inevitable as a basis for understanding human action) does represent a distinct methodological attitude or persuasion,[34] and one that is potentially incisive and very valuable in the study of counter-terrorism.

None of this means that there are simple lessons from history in this field. But close historical readings (when taken together with the insights from other approaches) can provide contextually informed intuition, analysis, explanation, perspective, and understanding that are of potential value regarding contemporary events and how we should respond to them.[35] It is hoped that this work can therefore helpfully complement what people from very different perspectives will bring to this important debate.[36]

IV

Does Counter-Terrorism Work? offers three case-study analyses (respectively, on the post-9/11 War on Terror, counter-terrorism in Northern Ireland,

and the case of Israel/Palestine), followed by a more wide-angled Conclusion. It is intended to balance the specific and the comparative. These particular sites of focus reflect numerous significant elements needing to be addressed in relation to our question (embodying, as they do, considerable variation in relation to geographical location, to scale, to the nature of the states under scrutiny, to the ambitions and dynamics of the terrorisms involved, and to the patterns of counter-terrorist policy and practice that have been adopted). The book's chapters all reflect my layered framework regarding counter-terrorist effectiveness, and my historically minded approach. *Does Counter-Terrorism Work?* also aims to offer insights about why some forms of counter-terrorist activity work better (or worse) than others, by highlighting issues such as the setting of realistic goals, the approach towards the root causes behind terrorism, the deployment of military means, the state's approach towards intelligence work and towards the law, questions of coordination within and between states, and also issues of credibility in counter-terrorist argument.[37]

I have addressed in other books the implications of the (in)efficacy of terrorism, the complexities of the relationship between terrorism and war, and the importance of understanding the mutually shaping relationship between counter-terrorism and terrorism.[38] This current volume complements those earlier arguments by trying to address, rigorously and calmly, one of the most analytically and practically significant questions of our era. No book, of course, will answer a question such as *Does Counter-Terrorism Work?* to the satisfaction of every reader. Rather, this study is intended to stimulate reflection and debate by offering an innovative, systematic argument about what *working* would mean in relation to counter-terrorism, an historically minded assessment of how far such successes have actually been achieved in practice, and some suggestions about the practical implications of these findings. This subject is of such importance, and our individual and collective responses to it matter so greatly in terms of shaping world politics, that we need an inclusive and honest debate about the issues raised in this book. If our collective response to terrorism is to be more effective and life-saving in the future than it has often been in the past, then we require dialogue on this vital subject between people from diverse backgrounds, professions, regions, interests, and political commitments. If *Does Counter-Terrorism Work?* stimulates debate (and perhaps fruitful disagreement) within and between such groups, then it will have served its purpose well.

PART
I

The Post-9/11 War on Terror

I

Afghanistan

I

Prior to the twenty-first century, the leaders of the United States on occasion considered themselves to be at war with terrorism. In the mid-1980s, President Ronald Reagan and key members of his team explicitly characterized the struggle of the US against terrorism as a war.[1] Moreover, some of the difficulties potentially involved in a war on terrorism were evident during the Reagan administration. Consistently matching US policy in practice to the ambitious promise of rhetoric proved impossible; and the divisions of opinion within the administration represented a frequent problem in terms of the coordinated maintenance of coherent goals.[2] Each of these aspects of historical experience might have been judged relevant to those in power in Washington from 2001 onwards.

Before Reagan's presidency, in the early 1970s, President Richard Nixon had considered revolutionary terrorism a major threat to the US, and there had indeed been many instances of anti-state terrorist violence in that period.[3] President Nixon urged the Federal Bureau of Investigation (FBI) to develop counter-terrorist tactics and capacity, and the Bureau did come to see terrorism as a major priority for US national security. Indeed, the Nixon years could helpfully be read through the lenses of our fourfold framework. At the tactical-operational level, there were considerable tensions between different wings of the state (particularly between the White House and the FBI) in pursuit of counter-terrorist endeavours. There was some tension between tactical successes and strategic goals, as counter-terrorist ventures helped to generate some responsive terrorist violence in escalatory cycles, and also to undermine public support for organizations such as the FBI. There were also inherent political rewards being sought by a president who stressed his commitment to restore law and order, and who sought to 'get these bastards!'. And there was considerable evidence of terrorist groups doing tactical damage to themselves through counter-productive violence and factionalism.[4]

Despite these illuminating earlier episodes, however, the post-9/11, US-led War on Terror[5] embodied an unprecedentedly vast effort to deal with non-state terrorism. More than two decades after the start of these efforts, it still represents a key locus for anyone assessing the effectiveness of counter-terrorism.

Terrorism was a comparatively limited priority for the US government immediately prior to 9/11,[6] and the atrocity of September 2001 prompted something of a sea change. For some political leaders, indeed, 9/11 really had changed the world, introducing a new kind of enemy, a dangerously novel terroristic threat, and the beginning of a new epoch in global politics. On 17 September 2001, President George W. Bush observed: 'I know that this is a different type of enemy than we're used to. It's an enemy that likes to hide and burrow in, and their network is extensive. There are no rules. It's barbaric behaviour. . . . And we're adjusting our thinking to the new type of enemy.' Just the day before, the Central Intelligence Agency (CIA) Director George Tenet had written that 'all of the rules have changed'.[7] Echoing this, former FBI Director James Comey said about 9/11: 'That day changed our country and it changed the lives of all of us in government.'[8]

This was not an inevitable reaction, and nor was it merely a US perspective. Tony Blair (in 2001 the Prime Minister of the United Kingdom, and one of the keenest allies of the US in the War on Terror) considered 9/11 to be:

> an event like no other . . . not an ordinary event but a world-changing one . . . it was accepted that the world had changed. . . . It was war. It had to be fought and won. But it was a war unlike any other. . . . This mass terrorism is the new evil in our world. . . . I conceived of September 11 as making all previous analyses redundant. . . . Even more so than those like me on the outside, those inside the American administration were clear: we had to take a wholly new look at the world.[9]

Even if one judges such analyses to exaggerate the extent to which terrorism and counter-terrorism had necessarily changed with 9/11,[10] it is important to recognize how widespread such interpretations have been, and also to recapture the explicable panic and anxiety of that moment immediately after the September 2001 attacks. Complex historical context is important here. Many people feared that other, perhaps frequent, terrorist attacks of this kind might follow. We now know that this did not happen, and that 9/11 represented al-Qaida's callous high-point of tactical terrorism, rather than the inauguration of a new phase of possibly even worse atrocities against the US. But many were not confident of that at the time. In the context of post-9/11 uncertainty, it was easy enough to overstate the danger: imperfect information prevailed, there was an understandable psychological anxiety, and many feared that they had entered a newly devastating period of mass-casualty terrorist attacks.

The US-led War on Terror consequently emerged soon after the atrocity. Speaking on 11 September 2001 itself, President Bush stated:

> Today, our fellow citizens, our way of life, our very freedom came under attack in a series of deliberate and deadly terrorist acts.... Thousands of lives were suddenly ended by evil, despicable acts of terror.... America and our friends and allies join with all those who want peace and security in the world, and we stand together to win the war against terrorism.[11]

What, though, were the aims of the war? During the post-9/11 US-led War on Terror there in fact remained considerable imprecision about what *winning* would actually mean.[12] But some crucial contours can be detected and clarified.

At times, it was clear that strategic victory was the criterion for the war's success. George W. Bush's famous comments of 20 September 2001 offer one important example of this: 'Our war on terror begins with al-Qaida, but it does not end there. It will not end until every terrorist group of global reach has been found, stopped, and defeated.'[13]

Much of the focus was on particular adversaries, especially al-Qaida and later ISIS (the Islamic State of Iraq and Syria). But the rhetorical ambition was set even higher than the destruction of these groups. In the period after 9/11, it was claimed:

> The United States, together with our Coalition partners, has fought back and will win this war. We will hold the perpetrators accountable and work to prevent the recurrence of similar atrocities on any scale—whether at home or abroad. The War on Terror extends beyond the current armed conflict that arose out of the attacks of September 11, 2001, and embraces all facets of continuing US efforts to bring an end to the scourge of terrorism.[14]

At times, however, some of what was sought in the War on Terror seemed to align more with partial strategic victory, and with a *substantial* but more limited ambition. President Bush's words sometimes, for example, implied that full victory might prove less likely than would a near return to normality, a prevention of terrorist victory, and a substantial but not complete counter-terrorist success: 'It is my hope that in the months and years ahead, life will return almost to normal. We'll go back to our lives and routines, and that is good.'[15] Partial strategic success could be claimed if terrorist adversaries were not destroyed, but significantly damaged and degraded in their capacity. Partial strategic victory might also be evident if the West could ensure the dominance of its own, pro-democratic narrative. As President Bush said to the US Congress in September 2001:

> Americans are asking, why do they hate us? They hate what we see right here in this chamber—a democratically elected government. Their leaders are self-appointed. They hate our freedoms—our freedom of religion, our freedom of speech, our freedom to vote and assemble and disagree with each other. They want to overthrow existing governments in many Muslim countries, such as Egypt, Saudi Arabia, and Jordan. They want to drive Israel out of the Middle East. They want to drive Christians and Jews out of vast regions of Asia and Africa. These terrorists kill not merely to end lives, but to disrupt and end a way of life. With every atrocity, they hope that America grows fearful, retreating from the world and forsaking our friends. They stand against us, because we stand in their way.[16]

Other presidential claims by President Bush pointed towards the same partial strategic goal of determining the narrative and agenda: 'We face enemies that hate not our policies, but our existence; the tolerance of openness and creative culture that defines us.'[17] Again, as set out in its 2006 iteration, the US strategy for the War on Terror saw it as 'both a battle of arms and a battle of ideas. Not only do we fight our terrorist enemies on the battlefield, we promote freedom and human dignity as alternatives to the terrorists' perverse vision of oppression and totalitarian rule'; the strategy 'involved destroying the larger al-Qaida network and also confronting the radical ideology that inspired others to join or support the terrorist movement'.[18]

Preventing enemy victory would also be judged a partial strategic success. In terms of al-Qaida, for example, the central, primary, strategic objectives for the US to thwart were the overturning of what al-Qaida considered apostate Muslim regimes, the expulsion of the US and US influence from Muslim countries, and the renaissance of a particular kind of Islam in tune with Salafist thinking (thinking which favoured the establishment of an order within which sharia law would be strictly implemented).[19]

The US-led War on Terror also possessed some clearly articulated tactical objectives. In 2019, the US Department of Homeland Security set out a framework of four key goals regarding terrorism:

> Goal 1: Understand the evolving terrorism and targeted violence threat environment, and support partners in the homeland security enterprise through this specialized knowledge. Goal 2: Prevent terrorists and other hostile actors from entering the United States, and deny them the opportunity to exploit the Nation's trade, immigration, and domestic and international travel systems. Goal 3: Prevent terrorism and targeted violence. Goal 4: Enhance US infrastructure protections and community preparedness.[20]

Similarly, in 2006 there had been a tactical commitment to 'prevent attacks by terrorist networks', 'deny weapons of mass destruction to rogue states and terrorist allies who seek to use them', 'deny terrorists the support and sanctuary of rogue states', 'deny terrorists control of any nation they would use as a base and launching pad for terror', 'deny terrorists entry to the United States and disrupt their travel internationally', and 'defend potential targets of attack'.[21]

Less publicly or frequently articulated, but still of great significance, were the ambitions of those who sought or identified inherent rewards through the War on Terror (whether in terms of popularity, fame, career advantage, financial gain, or psychological and emotional rewards).

How well were these various kinds of counter-terrorist success realized in practice? President Bush implied in January 2002 that victory was coming: 'we are winning the War on Terror.'[22] I'll now consider that crucial theatre of post-9/11 US counter-terrorism, Afghanistan from 2001 onwards.

II

The twenty-first-century Afghanistan and Iraq wars are too often blurred into one in popular memory. In fact, they had many differentiating dynamics, and the complex particularity of each context is crucial to understanding how the War on Terror developed in these respective settings. In the immediate aftermath of 9/11, the fact that al-Qaida had relied heavily on their Afghan base during the period of the attack's planning made it logical for the US government to act against them there and to deal also with the possibility of an enduring threat enabled by their Taliban hosts. While other regions of the world represented very sharp challenges with regard to al-Qaida and to terrorism more broadly,[23] it was in Afghanistan that the US decided initially to commit its most concentrated energies in the War on Terror.

The counter-terrorist goals of the US and its allies in Afghanistan during the 2001–21 period were fluid rather than utterly constant. But they can be usefully categorized in order to evaluate success. The central strategic aims were: pursuing, finding, capturing, killing, damaging, degrading, defeating, and destroying al-Qaida, and overthrowing the Taliban regime which had provided haven for al-Qaida. The latter was a major point. Understandable concern existed after 9/11, both about there having been a safe haven for those who had organized the merciless 2001 attack on the US, but also that such a haven should not exist in the future. Indeed, one of the main reasons for the post-9/11 US presence in Afghanistan lasting so long was this very concern: that terrorist attacks might originate again from that country.[24] Afghanistan was therefore one part of the wider struggle against terrorism and terrorism-supporting states, and part of the campaign to destroy international terrorism and its threat to the US.

Across the regimes of Presidents Bush, Obama, and Trump it remained a counter-terrorist objective to weaken al-Qaida and other terrorists in Afghanistan so that they could not again threaten the US. Disrupting al-Qaida, reducing their ability to attack the US, killing some of their leaders—such work might not destroy the organization, but it might still yield the partial strategic success of substantially limiting their capacity. Part of this reflected the notion of attacking terrorists abroad so that they could not attack the US at home. And part of it also involved the issue which remained a preoccupation from Presidents Bush to Biden: the desire to limit the likelihood that Afghanistan might again become a sanctuary from which terrorists hostile to the US could launch their lethal attacks.

Partial strategic success might also involve achieving something which had initially been a secondary goal, but the significance of which had grown during the post-2001 US involvement in Afghanistan. This was the aim of establishing a new kind of country, one in which greater democracy, liberty, equality, and opportunity might exist within a reconstructed society. This could serve the purpose of maintaining a Kabul government which would not in future host anti-US terrorists. But this secondary objective also possessed an intrinsic and idealistic quality in itself.[25] The regimes of Presidents Bush, Obama, and Trump all sought a peaceful, stable, secure, and well-governed Afghanistan, effectively a process of ambitious nation-building.

For such strategic and partial strategic goals to have a chance of success, tactical work in Afghanistan needed to be effective. At this level, there was the pursuit of operational successes, aimed at undermining al-Qaida's (and their Taliban hosts') capacity, and to punish those who attacked America by bringing them to justice. On the outbreak of Operation Enduring Freedom (the name given to the US's post-9/11 campaign against terrorism), President Bush thus declared that, 'the United States military has begun strikes against al-Qaida terrorist training camps and military installations of the Taliban regime in Afghanistan. These carefully targeted actions are designed to disrupt the use of Afghanistan as a terrorist base of operations, and to attack the military capability of the Taliban regime.... This military action is a part of our campaign against terrorism.'[26] Tactical successes might also involve good publicity about the righteousness and triumphs of the war, control over the relevant population, the day-to-day undermining of opponents on the ground, and the strengthening of the organizational infrastructure of both the US in Afghanistan and the reconstructed Afghan national institutions.

Strategic and tactical motivations need not preclude the pursuit of inherent rewards, however. One goal here, explicable in the wake of the terrible 9/11 attacks on the United States, was a desire for revenge against its perpetrators. As one US Special Forces officer starkly put it, 'Most of us wanted retribution', being keen on 'avenging the 9/11 attacks by putting bullets through al-Qaida and the Taliban'.[27] To hit back vengefully at the organization which had assaulted America, and at the regime which had hosted them as they planned it, offered an alluring prospect in the counterterrorist campaign of 2001 and subsequent years.

How did all these Afghan-related goals fare in practice?

Strategically, al-Qaida was not completely destroyed or eradicated[28] and, as we will see, the initial removal of the Taliban from power was eventually followed by their return to authority twenty years later. But full strategic success is rare across many areas of human endeavour, so could a case perhaps be made for substantial partial strategic successes having been secured by the US and its allies in post-2001 Afghanistan? The achievements were certainly flawed, but they were far from trivial.

The US engagement in Afghanistan did significantly damage al-Qaida, and the removal of the Taliban regime for a period made the safe-haven threat less dangerous. US-led action in Afghanistan made it more difficult for al-Qaida to operate there after 2001, with the Taliban support base uprooted for years, and with al-Qaida attacked, challenged, and degraded. Operation Enduring Freedom began on 7 October 2001 with US and allied forces quickly able to overthrow the Taliban militarily in just over two months from that date. This was a strikingly impressive success.

But a lengthy insurgency ensued, and blood-stained chaos marred Afghanistan for years after 2001. Likewise, deep societal difficulties endured. Opium production remained at high levels nearly two decades after the 2001 invasion, and violence was tragically ongoing: in 2017 alone there were over 10,000 civilians killed or injured from the hostilities.[29] Moreover, America's partial strategic success in Afghanistan has to be considered against the eventual re-emergence of the Afghan Taliban. By the end of 2016 much of the country was again under Taliban control and, as of March 2021, the Taliban (who still retained strong al-Qaida links) were in substantial control of significant sections of Afghanistan; at that stage the US-backed Afghan government's authority extended only to about a third of the country, and US forces in Afghanistan had fallen to around a mere 3,500.[30] Peace and stability had not been achieved as planned.

It is essential to assess counter-terrorism over long time-frames. While the future is contingent and unknowable, the 2021 US withdrawal from Afghanistan, together with the return to power of the Taliban in that year, significantly affects any serious assessment of this aspect of the War on Terror. US disengagement itself had very long roots. There had been a gradual shift of emphasis towards thinning out America's Afghan commitment, with a smaller presence, and a greater stress on indigenous forces being the guarantor of stability. On 28 December 2014, President Barack Obama declared the end of Operation Enduring Freedom in Afghanistan; by the end of that year most US troops had been withdrawn from the country. Under President Donald Trump a peace process was launched in 2018; in 2020 the US and the Taliban signed the Doha agreement, a deal which set a timeline for full withdrawal of US and international forces, anticipated to occur in 2021. Trump's successor (President Joe Biden, a long-term sceptic about the war in Afghanistan) duly delivered, deciding on a rapid withdrawal of forces from the country. On 14 April 2021, Biden announced that final US military withdrawal would be complete by 11 September that same year.

President Biden argued that the Afghan mission had had two goals, namely to 'get' Osama bin Laden (who had indeed been killed in 2011), and to prevent Afghanistan from being a base out of which terrorists could attack the US. He claimed that, these goals having been achieved, it was now time for the Afghans to do the work on their own.

In August 2021 President Biden spoke about 'how to protect our interests and values as we end our military mission in Afghanistan'. US personnel would imminently be leaving that country, he said, and the twenty-year mission there had seen key goals achieved: 'the death of Osama bin Laden' and 'the degradation of al-Qaida'. Further engagement was therefore not justified:

> Over our country's twenty years at war in Afghanistan, America has sent its finest young men and women, invested nearly one trillion dollars, trained over 300,000 Afghan soldiers and police, equipped them with state-of-the-art military equipment, and maintained their air force as part of the longest war in US history. One more year, or five more years, of US military presence would not have made a difference if the Afghan military cannot or will not hold its own country. And an endless American presence in the middle of another country's civil conflict was not acceptable to me.[31]

In practice, however, the US withdrawal unquestionably represented a setback in terms of the War on Terror and its objectives. One of the main US goals had been to remove the Taliban and thereby ensure that there would be no anti-Western terrorist safe haven; with the Taliban's return, this goal was put at very serious risk. Without ongoing US backing, the Afghan government and its regime collapsed rapidly and the Taliban returned to power, very quickly taking control again in 2021. US misjudgements about the likely course of post-withdrawal events made a contribution here, as did the corruption of those Afghan forces and political actors supposedly embodying an enduringly non-Taliban infrastructure in Afghanistan.

Some put a positive interpretation on this aspect of the War on Terror. As the final part of the American military evacuation took place in August 2021, the head of US Central Command, General Kenneth McKenzie, commented that:

> Tonight's withdrawal signifies both the end of the military component of the evacuation, but also the end of the nearly twenty-year mission that began in Afghanistan shortly after September 11, 2001. It is a mission that brought Osama bin Laden to justice along with many of his al-Qaida co-conspirators. The cost was 2,461 US service members and civilians killed and more than 20,000 who were injured. . . . We honour their sacrifice today as we remember their heroic accomplishments.[32]

But many Afghans were disappointed with what the US had achieved in the campaign, and the return to power of the Taliban offered (in addition to the suffering of ordinary citizens) the possibility that the partial strategic goal of removing terrorist safe havens might not turn out to have been achieved. After 9/11, the US and its allies had forcibly pushed the Taliban out of Afghanistan; now, twenty years later, the Taliban were back in control of the country, and militant-authoritarian Islamist rule had returned.[33]

Two aspects of this are especially worth considering.

The nation-building goal was one that had seen significant achievements. During the US engagement, there had indeed been many positives: there were greater rights for women and more freedom of the press; large numbers of Afghan refugees had returned to the country; more children had been educated; many new health care and other social service facilities and opportunities had been created; life expectancy had increased significantly.[34] But the 2021 return to power of the Taliban put *all* of these developments profoundly at risk, and reflected the seemingly long-term failure of the nation-building mission in the country. Any claims of even partial strategic success were undermined by the facts of large-scale migration and humanitarian crisis.

In relation to terrorism itself, there was also the problem that this Afghan experience might suggest (in a simplified and misleading manner) that terrorism does indeed work. That misrepresents the broader historical experience (in which terrorist organizations tend overwhelmingly not to secure their central goals).[35] But easily memorable, atypical examples can nonetheless affect wider judgements. Just as Osama bin Laden had been convinced that anti-Soviet jihadism in Afghanistan gave hope for terrorist victory against another super-power, so too some will draw the lesson that Taliban violence after 2001 in the end did show the efficacy of anti-Western terrorism. It may be that the new regime in Afghanistan ultimately decides that the costs of supporting anti-US violence are unappealingly high. If so, then the demonstration that enmity with the US would carry severe risks could be claimed to have produced partial strategic success. But the 2021 imbroglio (presented by the Taliban as a victory over the US) might yet lay the foundation for more terrorism. It would be wrong to ignore or dismiss the positive achievements secured by US and allied counter-terrorist efforts in twenty-first-century Afghanistan. Viewed over the long time-frame essential to serious-minded analysis, however, the 2021 return of the Taliban calls these achievements into doubt for the future.

Facilitating these partial strategic achievements (precarious though they currently may seem), there lay much tactical-operational counter-terrorist work and success. The skill and efficiency involved in the rapid defeat of the Taliban (within three months during late 2001) represented a major tactical-operational achievement for the US-led effort,[36] and ensuing years saw many other operational successes against both the Taliban and al-Qaida.[37] Huge costs were paid, however, in terms of US and Afghan people killed and wounded.[38] Moreover, it appears that Osama bin Laden himself narrowly escaped US pursuit at the end of 2001,[39] and the failure to capture or kill him for a decade—until his death on 2 May 2011—represented a tactical-operational setback in the struggle against al-Qaida.[40]

This had implications for public perception of the campaign, as did the profound damage—likely to be trans-generational damage—that was done to US credibility because of the collateral damage inflicted on Afghan civilians by American violence.[41] Credibility is a vital asset in effective counter-terrorism, as is the avoidance of over-reliance on military methods:[42] the unintended death and maiming of so many Afghan people in this military campaign severely undermined support for the US or the chances of America plausibly seeming to be defenders and allies of the country.[43]

At the tactical-organizational level, huge expenditure and effort went into trying to produce resilient infrastructure, since US counter-terrorism in Afghanistan depended not only on America's own capacity, but also on the strength and efficacy of Afghan institutions and their ability to resist the recrudescence of a Taliban challenge.[44] But part of the problem faced with the post-9/11 violence that endured in Afghanistan was that too little attention had been paid during the initial period after 2001 to the development of an effective Afghan police force.[45] Though huge amounts of money were spent, there remained insufficient resources, equipment, and people, and the Afghan military and police infrastructure was therefore not strong enough when the crisis arose.[46]

It was not that the importance of this factor was unforeseen. Karl Eikenberry (US and coalition commander in Afghanistan, 2005–7; US Ambassador to Afghanistan, 2009–11) had emphasized the need for a viably professional Afghan army, something which he aimed to create and which he hoped might provide the basis for responsible US withdrawal.[47] But the speed and extent of the 2001 military victory seemingly led to a US under-estimation of how strong an Afghan state and army and police would be required. In practice, there was not enough effective US planning for creating a resilient new Afghanistan.

Moreover, Afghan forces remained handicapped by corruption and mismanagement and thus had less chance of durability once US forces eventually departed. This was a crucial issue. The indigenous dimensions of counter-terrorism (local knowledge of the population, terrain, politics, and culture) can be vital in lending organizational strength, and so local security forces are of huge significance. In 2001, the US military had lacked sufficient intimacy of understanding regarding Afghanistan (the eye having been taken off the Afghan ball following the Soviet defeat there and 1989 withdrawal).[48]

III

The Afghan picture is therefore a complex and painful one. How can this set of outcomes be explained? No single reason will suffice. But the point about intimate knowledge is a necessary starting place. Long-term, historical reflection on Afghanistan would have highlighted the past experience of external forces who had enjoyed early success in Afghan campaigns only to find the eventual outcome of the war painfully negative; it would also have pointed to the past experience of leaders who provoked resistance with which their regimes could not cope.[49] But many of those involved in the US-led effort in Afghanistan had been ignorant of (and too often continued to ignore) the complex details of Afghan cultural realities and inheritances. The resulting overconfidence was one element explaining US failures after 2001.[50] The American-led effort in Afghanistan suffered from a lack of sharp-eyed understanding of the complex particularity of Afghan context, something reinforced by a failure of coherent and consistent long-term planning. This further contributed to the Afghan government and administration remaining impaired through corruption, inefficiency, and nepotism.[51]

It is also true that Pakistani and other (including Iranian) support for the Taliban made a difference to their strength, not least in terms of Pakistani safe havens for Taliban fighters and leaders. As noted above, collateral damage from US and allied engagements lost some hearts and minds, and the US-supported Afghan regime was also in many instances corrupt. I think it is also clear that the decision to focus on Iraq made it more difficult for the US to deliver on its goals in Afghanistan, diverting attention, expertise, key personnel, and resources from that more important counter-terrorist context.[52] Fighting on two major fronts from 2003 onwards made things considerably more stretched, and the damage done by the Iraq War to US credibility in counter-terrorism was huge (after the discrediting of so many aspects of the American justification for war in Iraq, the blowback generated by mistreatment of prisoners there, and the large-scale collateral damage done to civilians in the country).[53] Afghanistan slipped down the list of priorities, and the Iraq-based distraction and self-delegitimization proved extremely pernicious for the War on Terror.

Crucial too was the fact that the Taliban persuasively presented themselves as fighting for and defending Islam, a religion which represented such an historically deep-rooted and pervasive aspect of Afghan identity and experience. They cast themselves as offering resistance and defence against foreign occupation by an illegitimate, non-Islamic power, and as authentically and resonantly standing for what it really meant to be Afghan; and they powerfully took on the role of striking back in revenge for violence committed against Afghans by a foreign power.[54]

Religiously tinted nationalism; the politics of national legitimacy; deeply rooted resistance to foreign occupation and externally backed government; the desire for revenge in light of violence against one's own community—these crucial aspects of explanation are familiar (or should have been) from many other settings;[55] between them, these themes provided the basis for the intense Taliban commitment to the struggle. An historical and cultural understanding of Afghanistan would have pointed clearly (in advance) to the likely importance of such themes; but such understanding was largely absent from US policy-making when 9/11 reignited American interest in the country.[56] Historical and cultural understanding of a country does not of itself guarantee successful foreign policy; but it might reasonably be judged one essential component behind successful engagement.[57]

The Taliban's own actions, of course, also contributed significantly to the 2021 outcome. Effective US counter-terrorism in Afghanistan would have involved more than merely military force. It would have required winning a war of narratives, conducting successful information operations, and making clear the advantages to Afghans of dissociating themselves from the politics of al-Qaida and the Taliban alike. In practice, the Taliban were in fact winning this form of war against the US and its allies. Excellent at the weaponizing of narratives within politics, their message and their means of communicating it resonated far more successfully with Afghans than did those of their US opponents. Culturally resonant themes, references, priorities, values, and modes of communication were eschewed by the US; the process of securing trust, support, and confidence from sufficient numbers of Afghans was damaged as a consequence.[58] In 2017, several years before the US withdrawal, Afghan expert Thomas Johnson perceptively wrote that:

> The US and NATO would have been well advised during the initial stages of Operation Enduring Freedom to ask simple questions about what Afghans—not just the elites in Kabul, technocrats, and expatriates—really wanted from the West's engagement. What are the messages and frames that motivate and influence Afghans? What is the best way to spread a message? Such questions were never systematically asked.... This lack of understanding, in part, has ultimately doomed Western engagement in Afghanistan and contributed to the West losing the battle of the story in Afghanistan and, therefore, the war.[59]

In reflecting on the 2001–21 Afghan elements of the War on Terror, it is vital to think about the long-rooted dynamics involved. For Western engagement with the country had long involved complex outcomes in relation to terrorism. During the period of late twentieth-century Soviet influence in Afghanistan, for example, the US had supported rebels against their Cold War enemies. On 25 December 1979, the Soviets had invaded the country; the CIA subsequently directed funding to the anti-Soviet Mujahideen during the 1980s via Pakistan's Inter-Services Intelligence (ISI); in 1992 Kabul did indeed fall to the CIA-backed Mujahideen. While the US did not fund Osama bin Laden in Afghanistan, some of those involved in the rebel violence against Soviet occupation were indeed proto-al-Qaida zealots; and bin Laden (who was himself involved in the anti-Soviet jihad) did draw the lesson that, just as violence here had in his view helped to undermine one super-power, so too future terrorism might undo the US: 'We believe that America is weaker than [Soviet] Russia', as bin Laden himself put it.[60]

However misguided his assessment (and bin Laden overplayed both the role of the Arab volunteers in the Mujahideen struggle,[61] and also the role of Afghanistan in destroying the Soviet Union), the issue here concerns the power of perceptions. In those terms, the US's 1980s engagement in Afghanistan had unintentionally contributed to the intensification of anti-Western terrorism, with appalling consequences on 11 September 2001. Part of this trajectory was also a failure to recognize that the most effective time to deal with terrorism is actually before it becomes a famous and dangerous threat. But between the Soviet departure from Afghanistan in 1989 and the 9/11 atrocity in 2001, Afghanistan was allowed to become a harbour for al-Qaida, with far too little Western intelligence work being done in the country.[62] In the late 1990s, for example, the CIA did not have even one case officer capable of speaking Pashto (the language of the major ethnic group in Afghanistan).[63] Just as there were clear intelligence failures in the direct run-up to 9/11 itself,[64] so too there was therefore a broader context of intelligence myopia in regard to this crucial Afghan setting.

Another insight from history is the recognition that recent Afghan failures possess broader regional dynamics; this again contributes to an understanding of how the counter-terrorist effort in the country has fared since 2001. In particular, political and security dynamics relating to Pakistan have been and remain of high significance for Western counter-terrorism. Afghanistan and Pakistan share a long border, and the Pakistani government clearly has a profound interest in who rules their neighbouring country.[65] Pakistan's ISI gave assistance to the Taliban in the 1990s and early 2000s, with a view to having a powerful ally in Kabul[66] and a government there that was not sympathetic to India. Seeing their geopolitical interests to be advanced by offering such support, Pakistan (ostensibly a post-9/11 ally of the US) has in fact also played an ongoing role in countering American goals in Afghanistan. A country with a long record of using militant proxies, Pakistan offered sanctuary beyond the Afghan border to some people whom the US was now pursuing. By 2019, about twenty terrorist groups were functioning in Afghanistan and Pakistan.[67] It is important here to recognize both the heterogeneity of Pakistani terrorism (which extends far beyond the famous jihadist variety), and the fact that most victims of Pakistan-related terrorism are Pakistanis themselves.[68]

Does historical respect for contingency suggest that things might have turned out differently in the West's post-2001 Afghan venture? Some will see an inevitability to the traumatic 2021 ending of this US endeavour. But it is important to remember how much Afghan support there had been in 2001 for initial US engagement,[69] and there were numerous aspects of the subsequent twenty years that might, perhaps, have developed differently. Without the Iraq distraction, could things have been done more successfully in Afghanistan? Should the Taliban have been included in the new political arrangement after the invasion had succeeded, with much earlier dialogue with them being pursued? As expert voices have pointed out, Western states have ended up talking with the Taliban; those same voices have at times suggested that earlier engagement with the Taliban would have been more propitious.[70] Indeed, many Afghans have shown a pragmatic approach towards alliances during conflict,[71] and this might have facilitated support for more positive engagement with the very powerful, then-victorious US at an earlier stage. Would it have been possible at that early phase of the conflict to reach an arrangement which precluded anti-US terrorism re-emerging from Afghan safe havens? And could that have been an arrangement without a long-term US presence in the country? Indeed, should the US have left Afghanistan in 2002, shortly after having bruised al-Qaida and overthrown the Taliban regime? Could more clear-sighted, realistic (and more consistently pursued) strategic objectives have generated a better outcome for the USA?[72] The nation-building, emancipatory aspects of the campaign in Afghanistan have suffered enormous setbacks. It therefore seems appropriate to ask: if the eventual outcome is indeed one in which the Taliban control Afghanistan, but perhaps ultimately choose not to support anti-US terrorists for fear of violence and reprisals from the US,[73] then could that not have been achieved far earlier, and without such vast loss of life, limb, money, and prestige?

Informed by a framework such as the one I'm adopting, US goals could perhaps have been more realistically established and more rigorously sustained with a sharper eye to what was actually achievable. Rather than allowing early-conflict tactical success to encourage exaggerated strategic ambitions, more appropriate tactical and partial strategic goals could have been identified and pursued. These would have included: the capture or killing of Osama bin Laden and his key colleagues, thereby significantly damaging al-Qaida; and the partial strategic aim of ensuring that neither the Taliban nor any other Afghan regime would host anti-Western terrorists (through demonstrating to them the painfully heavy cost to be paid for such behaviour). The best that can be claimed at time of writing is that al-Qaida were indeed damaged, and that it might eventually prove possible to dissuade the Taliban from offering sustained support and haven to anti-US terrorists. But achieving this latter aim seems far from certain, and in any case the cost has been extraordinarily and avoidably high. Because so much more was attempted (the ambition of producing a new kind of politics and society in Afghanistan), and because those attempts shambolically fell to the ground with the hasty and chaotic US and allied departure from the country,[74] the narrative of US defeat is one that will likely long endure. And the consequences of that credibility-damaging legacy could be very high indeed.

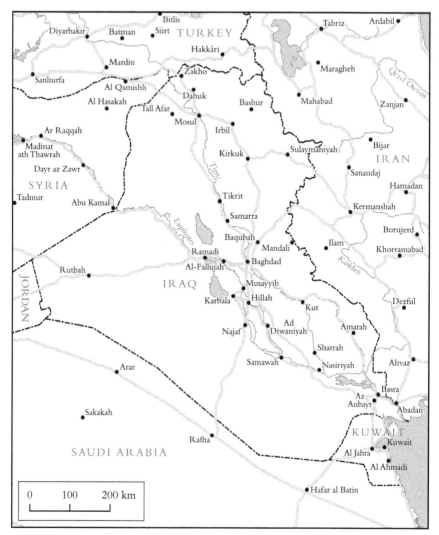

Map 1 Iraq

2

Iraq and ISIS

I

If Afghanistan represented one vital theatre for the War on Terror, then Iraq offers another unavoidable context for analysis. It is clear that US governmental focus turned to Iraq immediately after the 9/11 atrocity, that US Secretary of Defence Donald Rumsfeld and his Deputy Paul Wolfowitz strongly pushed this agenda, and that plans for the invasion of that country were submitted to President Bush before the end of 2001.[1]

The goals outlined for the Iraq War have become famous, and part of its ostensible rationale was to counter the supposed threat of terrorism generated by Saddam Hussein's regime and its connections. The overthrow of Hussein and his Ba'ath regime was a major war aim for the US-led coalition, and the reasons given for seeking such regime change were essentially threefold. It was claimed (first) that Hussein had links with al-Qaida, and that the terrorism already committed by bin Laden's organization (as well as the danger of possible further attacks) legitimated the removal of a state ally who could make terrorists even more dangerous to the USA in the future. Reinforcing this was the claim (second) that Hussein's regime possessed Weapons of Mass Destruction (WMD), and that the war in Iraq would allow for their removal. Anxiety about terrorists gaining WMD was articulated by President Bush emphatically in January 2003: 'Today, the gravest danger in the War on Terror, the gravest danger facing America and the world, is outlaw regimes that seek and possess nuclear, chemical, and biological weapons. These regimes could . . . give or sell those weapons to terrorist allies, who would use them without the least hesitation.'[2]

There was also the claim (third) that, given Saddam's brutality towards his own people, it was justifiable to replace such a heinous, dictatorial regime with one which would lead to greater freedom and democracy in Iraq. In terms of counter-terrorism, the major goals here therefore tended towards the strategic aim of removing a major threat (Saddam Hussein's Iraq as a terrorist sponsor) via the tactical means of military-operational success and the undermining of opponents (if al-Qaida, for example, could be denied logistical and other support from a supposed ally). The Iraq War was therefore tightly integrated into the War on Terror.[3]

How did the pursuit of these goals fare in practice?

The strategic aim of removing Saddam Hussein from power was skilfully and speedily achieved. As so often with counter-terrorism, however, matters were more complex in context. In the sense of removing a dictator from power, this achievement could indeed be judged a success. But the war had been justified partly because of Hussein possessing WMD (which he did not), and of his allegedly having been an ally of al-Qaida and a possible future threat as a sponsor of such groups on a catastrophic scale (claims that were also mistaken).[4]

Could a case be made that the Iraq War resulted in partial strategic counter-terrorist success? In terms of delivering diluted versions of central strategic goals, the answer must be negative. Since Hussein did not possess the weapons nor the alliances that had been central to justifying the war, the outcome here in counter-terrorist terms was simply negative. But what about the secondary goal of recasting the Middle East along more democratic lines? Could this also have involved partial strategic success in terms of determining the political agenda?

Here, the results were arguably more dismal still. Many problems with the Iraq War lay with the lack of effective planning for its aftermath; over-confidence had generated an unhelpfully casual approach. Indeed, in terms of combatting terrorism, the war proved counter-productive. This was partly because the basis for the invasion had included dubious claims about the Iraqi regime and al-Qaida, and partly because post-invasion Iraq became a magnet and a convenient justification for anti-Western terrorism. Far from leaving the Middle East in a more benign condition, the Iraqi chaos, the delegitimization of US credibility through the discrediting of American arguments, and the generation of a new site for terrorist energy all made the US situation worse.

Instead of emerging as a beacon for reform, post-war Iraq seemed to many to embody chaos rather than stability. The US-led attack impressively overthrew Saddam Hussein's regime in three weeks. But the subsequent imbroglio and lengthy conflict substantially undermined claims regarding the war aims' achievement. The US hope had been to establish democracy, economic transformation, and a thriving civil society. What was in fact produced was (in the words of one leading authority) 'a troubled and increasingly insecure country in which insurgency, lawlessness, and sectarian conflict claimed growing numbers of Iraqi lives, in addition to taking a mounting toll of the occupation forces'.[5]

Compounding this was the lasting damage done to US counter-terrorist efforts by the erroneous nature of crucial US claims that had been made to justify the war. Determining the agenda therefore became more difficult still. Again: the claims of alliance between Saddam Hussein's Iraqi regime and al-Qaida, and of his having had some responsibility for 9/11, were unfounded.[6] Hussein did not, in fact, have strong ties to al-Qaida, and President Bush was forced eventually to admit, on 17 September 2003, that: 'No, we've had no evidence that Saddam Hussein was involved with September 11.'[7]

It had also been claimed (by figures such as Vice President Dick Cheney and US Secretary of State Colin Powell, as well as by President Bush and Secretary of Defence Rumsfeld) that Iraq possessed significant WMD. It became clear after the Iraq War, however, that Saddam Hussein had not possessed the WMD that had been central to justifications for war against him. He had been much less of a threat than had been asserted; and the supposed need to prevent him from helping groups such as al-Qaida to use WMD likewise turned out to have been absent. The undermining of these arguments damaged US counter-terrorist credibility, domestically as well as internationally, for many years to come.[8]

More painful still was the reality that Iraq now became both a seeming justification for anti-Western violence and also an arena to which people could move to fight. 'You Americans are going to find that it is not so easy governing Iraq,' Saddam Hussein told one of his interrogators after his 2003 capture.[9] Indeed, the persistent theme of terrorism and counter-terrorism mutually shaping one another in blood-stained ways[10] found yet more expression in Iraq after the invasion. Collateral damage by the US in Iraq directly generated more anti-American violence, effectively more terrorism;[11] more broadly, the Iraq War stimulated more anti-Western, jihadist terrorism internationally (exactly the phenomenon that it was supposed to be eradicating).[12] Richard Clarke (admittedly, a Democrat, and so inclined towards scepticism about a Republican president) had been US National Coordinator for Security, Infrastructure Protection, and Counter-Terrorism during 1998–2001, and was to observe: 'George Bush was right...when he said that, "Iraq is the central front in the War on Terror." He made it so. He turned it from a nation that was not threatening us into a breeding ground for anti-American hatred.'[13]

Evidence does indeed show the effects of the Iraq War on terrorism to have been escalatory.[14] The CIA analyst who first extensively interrogated Saddam Hussein after his December 2003 capture stated that, 'Saddam's removal created a power vacuum that turned religious differences in Iraq into a sectarian bloodbath'.[15] Prominent counter-terrorists have been admirably candid at times about this outcome. As former Director General of the UK Security Service MI5, Eliza Manningham-Buller, crisply put it: 'Without doubt our involvement in Iraq spurred some young British Muslims to turn to terror.'[16] Nor was this surprising, given what terrorism's long past has involved. Revenge has long been a key motivation for terrorist recruitment, operations, and sustenance.[17] There also exists much historical evidence that the killing of civilians by states engaging in counter-terrorism can generate a backlash and strong support for the terrorism that is supposedly being extirpated.[18]

Hence, to sum up, terrorism actually grew as a consequence of the Iraq War: the country became a magnet for jihadists, a site of violent struggle and lasting chaos, a justification deployed in defence of terrorist attacks elsewhere, and part of the complex sequence of events which generated the ISIS crisis. US foreign policy had been one of the supposed justifications for al-Qaida violence for some years;[19] more broadly, Iraq fits the pattern that Western foreign policy—aimed at countering terrorism—has sometimes been one of the motivations behind (and one of the most persistent of justifications offered for) anti-Western terrorism.[20]

At the tactical-operational level, the Iraq War did witness some striking counter-terrorist successes. The early stages of the invasion were militarily successful. Operation Iraqi Freedom was established on 19 March 2003; the bombing of Baghdad commenced on 20 March; on 9 April coalition forces gained control of the city. Resistance and conflict were to follow. But even during those years there were many tactical-operational positives, as with the progress made during 2005–8 in al Anbar Province. Here, significantly, there had emerged greater interaction, trust, and positive engagement between Sunni tribes and US forces. The process involved both sets of actors reimagining each other, and seeing an opportunity for alliance where previously there had been a perception of threat.[21]

It would be mistaken, therefore, to dismiss the counter-terrorist effectiveness evident here, not least with the successful invasion and speedy overthrow of Saddam's regime. But, between then and US combat troops leaving Iraq in 2011, chaos in the country became endemic,[22] with criminality, insurgent resistance, terrorism, sectarianism, and lawlessness all increasing. Much of this reflected poor tactical achievements by the US. One crucial aspect here was the field of intelligence, so vital an element in effective counter-terrorism.[23] Pre-invasion, the CIA had been poorly prepared.[24] Intelligence failures had certainly contributed to the WMD fiasco. Both the US and their UK ally selected, interpreted, and used evidence in, arguably, dubious ways in order to justify and win support for the conflict.[25] The United Kingdom's Report on intelligence, WMD, and Iraq judged that, 'A major underlying reason for the problems that have arisen was the difficulty of achieving reliable human intelligence on Iraq', and found that there was 'no evidence of deliberate distortion or of culpable negligence'. But, in reference to the threat actually posed by Iraq, it concluded that 'there was no recent intelligence that would itself have given rise to a conclusion that Iraq was of more immediate concern than the activities of some other countries'. In regard to intelligence and the Iraq War, this Review judged that, 'Validation of human intelligence sources after the war has thrown doubt on a high proportion of those sources and of their reports, and hence on the quality of the intelligence assessments received by Ministers and officials in the period from summer 2002 to the outbreak of hostilities.'[26]

Once the war was under way, and the conflict had intensified with violent resistance to the occupation, the tactical arena of publicity was another area of US failure. The Abu Ghraib prisoner abuse scandal did serious damage to the reputation and therefore the counter-terrorist ambitions of the US,[27] reinforcing the problems that had already arisen after US-held detainees began arriving at Guantanamo Bay detention facility in 2002.

In terms of tactics and the securing of control over a population, there were clear successes in Iraq, but also many failures as the chaotic conflict evolved, and rival zones of authority challenged the US and its allies.[28] Despite contemporary and expert warnings,[29] there seems to have been a profound naivety on the part of the US, UK, and their allies regarding what would happen in post-invasion Iraq. Here, as so often, some knowledge of history might have been of great benefit. Careful consideration of the country's past would have demonstrated numerous vital points. It would have shown how difficult it has proved for those in authority in Iraq to achieve popular endorsement of their own objectives and preferences. It would have clarified how challenging it has been for those in political power to secure control over the country's diverse population. It would also have demonstrated how deeply embedded within Iraq has been a lasting culture of extreme violence.[30] So warning lights did exist, and thoughtful historical reflection might have helped. Moreover, given the lengthy involvement of the US and UK in Middle Eastern politics, there can be few plausible excuses for Washington and London not engaging in such historical reflection. Iraq's political origins and past experience strongly suggested, for example, that external invaders might face unexpected, unintended, and unwanted developments and responses; they also pointed to profound limitations in what might be anticipated to emerge from an invader's superior military force.[31] But the historian's (and others') emphasis on contingency, complexity, and long-term memory was over-ridden by other priorities and approaches, with catastrophic consequences for many people, and with damaging effects on Western counter-terrorism.

On the ground, day-to-day tactical failings contributed to strategic problems. In addition to the marked lack of planning for post-invasion Iraq,[32] there was too much optimism about what could be done, and a damaging underestimation of the difficulty, cost, and long-term timeframe necessary for post-war reconstruction, security, and stability.[33] All of this generated new problems for the already troubled region. The detailed planning that helped to make the military campaign so stunningly successful was not matched by long-term preparation for post-invasion life. Paul Bremer, US administrator of the Iraqi Coalition Provisional Authority during 2003–4, had no prior Middle Eastern experience; and his dissolution of the Ba'ath Party and proscription of its senior members from holding public service roles, together with his dissolution of the Iraqi armed and security forces, represented serious errors: experienced Iraqi administrators were now lost to administration, and thousands of armed young men became unemployed and dangerously hostile to the new order.[34] First-hand testimony from those involved in Iraq during this period reinforces not only the non-inevitability of the country's trajectory from 2003 onwards, but also the lamentable lack of planning in Western states' approach to Iraq's post-war development.[35]

In its counter-terrorist efforts, but more broadly too, the US was lastingly damaged organizationally by all these failures. The vast costs in lives, resources, and credibility left the USA less powerful as a state, and more frail and flawed.

The culture of counter-terrorism can involve, however, more than merely the countering of terrorism. What were the inherent rewards accrued from Iraq? There could be the fulfilling of a desire for revenge. One US Air Force major who flew post-9/11 special operations missions in Afghanistan and Iraq reflected on the satisfaction of 'getting a little bit of payback. Getting a little bit of personal payback.'[36] There could also be economic benefits. Private US contractors and companies made large amounts of money from the war. In 2010, it was discovered that there were almost 2,000 private companies in the US working on homeland security, counter-terrorism, and intelligence. During the Iraq War itself billions of dollars were flown from the US into the war region with very little supervision, process, or record-keeping to monitor what actually happened to the money; many of these dollars disappeared, and it seems that contractors were sometimes paid very significant sums for projects that were never in fact pursued.[37]

II

There are those who have offered careful defences of the Iraq War as a justified enterprise.[38] I myself opposed the war at the time in 2003, but I have long thought that the endeavour involved complexities rather than simple or self-evident assessments. My analysis has therefore sought to discern and acknowledge the successes that were indeed secured, while overall judging the venture to have been profoundly damaging in terms of counter-terrorism.

Could it have been different? Why did the mistaken judgements regarding Iraq occur? Partly, there was the issue of a particular political context. The 9/11 attacks seemed to some to call for a dramatic statement of American power globally; that context also offered seeming political rewards for such militaristic action. Partly, there were inherited elements, with the legacy from earlier US-Iraq engagement and its associated sense of unfinished business, together with a neo-conservative belief in determining the future.

Some have also pointed to explanations which are partly personal in their contingency. Philosopher Quassim Cassam has suggested that epistemic vices—'systematically harmful ways of thinking, attitudes, or character traits'—have had politically pernicious effects, including those evident in relation to the Iraq endeavour under discussion here. 'Epistemic vices get in the way of knowledge. They obstruct the gaining, keeping, and sharing of knowledge and it's because they do that that they can have disastrous consequences in the political realm.' In regard to the Bush regime's mis-preparation for the aftermath of the 2003 Iraq invasion, Cassam suggests that, 'Arrogance and overconfidence were two of the factors that caused Donald Rumsfeld and his colleagues to go so badly wrong in their thinking and planning. Arrogance and overconfidence are epistemic vices and the Iraq fiasco is an object lesson in how vices of the mind can obstruct our attempts to know things. . . . A list of the intellectual vices that contributed to the Iraq fiasco would also include dogmatism, closed-mindedness, prejudice, wishful thinking, overconfidence, and gullibility.' Cassam acknowledges that struc-tural considerations as well as individual ones will be significant when explaining complex phenomena such as the 2003 Iraq War;[39] but the contingent nature of the individuals and their failings that he sharply iden-tifies remains important, nonetheless.

Part of this fits a depressingly wider pattern of frequent state over-reaction to non-state terrorist provocation. In the wake of an appalling terrorist assault, the threat can become exaggerated, eclipsing more responsible attempts to keep the danger in proportion. In the post-9/11 US, the authorities frequently exaggerated the threat posed by jihadist terrorism,[40] with US Secretary of Homeland Security, Michael Chertoff (in post during 2005–9), even representing the fight against terrorism as 'a significant exist-ential one'.[41] In truth, terrorism did not and does not embody an existential threat to the US, and there already existed measures within the pre-9/11 United States to deal with most terrorism effectively.[42]

Whatever its origins, the conception of the Iraq War involved legitimiz-ing it as an essential part of the post-9/11 War on Terror. But one of its many unintended effects was to contribute to a complex sequence of events which led to the emergence of a new form of terroristic adversary in ISIS.

III

On 29 June 2014 the Sunni Muslim organization ISIS declared the estab-lishment of a substantial caliphate across parts of Iraq and Syria. In response, the US-led campaign against ISIS (the Combined Joint Task Force— Operation Inherent Resolve, established in 2014) had substantial international support, though it was emphatically a US-defined mission. It also contributed directly to the profound diminution and degrading of ISIS as a regional phenomenon. ISIS's loss of Raqqa in 2017 can be seen as effectively marking the end of the ISIS caliphate. Operation Inherent Resolve involved a combination of limited direct involvement (tactical operational work by Special Operations Forces teams, and through air strikes) with the deployment of proxy forces on the region's front line, and an engagement in cyber conflict.[43] The Iraq War and its chaotic aftermath had produced a Western reluctance to become too heavily or directly involved in military campaigns and commitments in contexts such as that now involving ISIS; so Operation Inherent Resolve occurred under the shadow of Operation Iraqi Freedom, and was somewhat risk-averse in nature as a result.

Let me stress again that in assessing post-9/11 counter-terrorism it is important to acknowledge how interwoven the emergence of ISIS was with the US-led response to the 9/11 attack. That 2001 attack prompted the US to engage in a vast counter-terrorist effort, part of which involved a war in Iraq which led to chaos and violent resistance. Amid that resistance, Al-Qaida in Iraq (AQI) emerged as a prominent actor, and it was out of this grouping that ISIS was to grow.[44] The US aim with ISIS was to degrade and destroy it, and the effective defeat of ISIS as a regional force was indeed substantially achieved by US-led efforts across the Obama and Trump regimes.[45] US troops were therefore able to complete their withdrawal from Syria during 2019.

Was ISIS completely destroyed? No: they remained a residual force and could on occasions be lethal as a network or brand, and as a medium for a particular ideological argument. In this way, ISIS retained something of a virtual presence after its physical base had been removed, and not all ISIS-related threats disappeared. Even as late as December 2020 the United Nations Analytical Support and Sanctions Monitoring Team noted that ISIS was still estimated to possess 10,000 'active fighters' in Iraq and Syria.[46] And contingencies could intervene. This same 2020 Report observed that, 'In conflict zones, the threat continued to rise as the pandemic [COVID-19] inhibited forces of law and order more than terrorists.' It also noted, however, that ISIS did not have a strategy for action during the pandemic.[47]

The Western campaign against ISIS had led to the terrorist group becoming so fragmented and substantially defeated that this aspect of post-9/11 counter-terrorism must be judged at least a partial strategic success for the US and its allies. Completely eradicating even one major terrorist organization is rare, and it should be recognized that ISIS was indeed strikingly degraded, its caliphate brutally ended, its financial foundations substantially demolished,[48] and the surge of foreign fighters to the region dried up. The partial strategic success of defeating it as a regional player and markedly diminishing it as a force had been secured. In September 2014, President Barack Obama had said that the anti-ISIS coalition aimed to 'degrade, and ultimately destroy' ISIS,[49] and this was substantially realized.

Part of this involved the dynamics of self-degradation as well as those of state hostility, as ISIS atrocities ultimately lost it significant support. It is also true that there existed less of a long-term US vision for what should happen after military victory in Syria and Iraq. A political solution to the problems which had created ISIS in the first place was less satisfactorily generated than were the tactical success and partial strategic victory alluded to above. The widespread sympathy that ISIS initially acquired owed much to explicable Sunni Muslim concerns in Syria and Iraq about the nature of government in those countries.[50] Until stability and legitimacy emerge in the region, the potential for future violent escalation is likely to remain.

3

Winning the War on Terror?

Afghanistan, Iraq, and the pursuit of ISIS all formed major parts of post-9/11, US-led counter-terrorism. Broadening our vision even further, what does a systematic overall assessment of counter-terrorism during the War on Terror reveal?

Strategic victory, in the ambitiously expressed goals of destroying al-Qaida and also international terrorism more generally, was always likely to be elusive, and so it has proved.[1] This is reinforced by reflection on the different kinds of violence that could variously be considered to be al-Qaida terrorism. There could be terrorist plots possessing a core al-Qaida role in terms of organizational command and control; but there could also be attacks more loosely proposed or endorsed by al-Qaida, or even ones that were self-initiated but somewhat inspired by the group.[2] Damaged though it has been, a less-coordinated al-Qaida still exists, its threat more complex, with inspiration and affiliation now being significant aspects of what it offers.

More persuasive claims can clearly be made for various kinds of partial strategic victory by the post-9/11 US and its allies. As noted, al-Qaida, ISIS, and other anti-Western terrorist groups have repeatedly been harmed, degraded, and undermined, leading to a reduction in terrorist capacity and a substantial but not complete counter-terrorist success.

Moreover, a realistic and important secondary goal of counter-terrorism, and an impressive achievement if secured, would be the reassurance that the state (and life within that state) were indeed to be maintained in ways that assured a continuation of normality, despite terrorist attempts to disrupt this. The damage done by jihadist violence had been intended to undermine normal societal confidence; in the month following the 9/11 atrocity, Osama bin Laden claimed, 'There is America, full of fear from its north to its south.'[3] And the immediate economic effects of the 2001 attack were profoundly damaging for the US.[4] But neither attitudinally nor economic-ally was the harm transformational in terms of daily life for most US citizens. Nor was the terrorist threat on such a scale that it would end the functioning of US society. Between 9/11 and 2015, jihadist terrorist attacks in the US killed nineteen people.[5] These were terrible tragedies but they certainly did not represent a major threat to the US as a whole.

It is impossible to ascertain how much of such normality was secured directly by counter-terrorist efforts, and some observers have been sceptical.[6] But there have been some major Western counter-terrorist successes, pre-venting attacks which would have involved loss of life. While the War on Terror was emphatically US-led, it did involve some allies, prominent among them the UK. The UK's 2018 iteration of its counter-terrorism strategy[7] affirmed that, 'CONTEST's overarching aim remains to reduce the risk to the UK and its citizens and interests overseas from terrorism, so that our people can go about their lives freely and with confidence.'[8] This would seem a more realistic and fruitful goal to pursue than promising the extirpation of terrorism.[9] Crucial to the formation of CONTEST was Sir David Omand, who had been Director of GCHQ (Government Commu-nications Headquarters, the UK's agency responsible for signals intelligence) during 1996–7, and who was Security and Intelligence Coordinator in the UK Cabinet Office during 2002–5. Omand's clear view of how counter-terrorism policy should be framed involved the following assessment:

> Expressing the aim in risk reduction terms avoids the corrosive effect of impossible promises, such as to eliminate terrorism, and leads naturally into framing campaigns that can reduce the overall risk.... There are limits to what government can reasonably do to protect the public. Care will continue to be needed not to fuel an illusion that life can ever be made risk-free.

Omand stressed that 'the power of normality' (whereby, despite the terrorist threat, the state ensured that people could pursue their normal life freely) denies terrorists 'part of what they most seek, which is to destabilize confi-dence in the authorities'.[10]

Close reading of the UK's formally declared approach does indeed dem-onstrate its emphasis on the need for a realistic and proportional response to terrorism:

> We will ensure our response reflects our guiding principles of proportionality, flexibility, and inclusivity. Terrorists attack us to create fear, to take revenge for

real and perceived grievances, and to influence public opinion. We will respond proportionately and in a way that does not undermine our aim to enable people to live freely and with confidence. The threat from terrorism will not go away and we may not always be successful in stopping attacks, but terrorists cannot and will not change our way of life.[11]

Such a stance resonates with that of former MI5 Director General Eliza Manningham-Buller, who commented in relation to the UK authorities' response to al-Qaida immediately after 9/11: 'What could we do to reduce apprehension and to encourage our citizens to continue their lives, as far as possible free from fear?'[12]

The UK's approach as embodied in CONTEST involved an alliterative sequence of objectives: 'Prevent: to stop people becoming terrorists or supporting terrorism; Pursue: to stop terrorist attacks; Protect: to strengthen our protection against a terrorist attack; Prepare: to mitigate the impact of a terrorist attack.'[13] There were continuities with pre-9/11 approaches here,[14] and also resonances with counter-terrorist policy elsewhere: the EU in 2005 set out to protect, prepare, pursue, respond; Australia's 2010 policy involved analysis, protection, response, resilience;[15] the US's 2003 Counter-Terrorism Strategy aimed at defeating terrorist organizations, denying terrorists the benefits of state sponsorship, diminishing the conditions and roots which generated terrorism in the first place, and defending the US and its interests through stronger security.[16]

The business of preventing terrorist attacks is clearly central to all this. As one former UK counter-terrorist police officer puts it, prevention is 'the main aim of governments as they not only try to keep their citizens safe, but also attempt to show that they have the capability to do so, thereby nullifying the effects of terrorism on their populations'.[17] Evidence suggests that there was indeed some success in ensuring that terrorist attacks did not end normal life in the post-9/11 UK in terms of actual threat. It seems, for example, that forty-three terrorist attacks or plots were thwarted by the authorities in the UK during 2001–12.[18] The scale of terrorist violence in the early twenty-first-century UK has not substantially damaged societal normality. The Office for National Statistics figures for England and Wales record ninety-three terrorism-generated deaths (excluding perpetrators) between April 2003 and the end of March 2021.[19]

In David Omand's words, CONTEST 'had a clear strategic aim: to make it possible for society to maintain conditions of normality so that people could go about their daily business, freely and with confidence, even in the face of suicidal terrorist attacks. The conditions, freely and with confidence, are an important reminder that we should seek security in ways that uphold our values such as liberty and freedom under the law.'[20] The disruption of terrorist plots goes some way to suggesting that this aim had been significantly achieved by the UK authorities. Indeed, for all of the fears understandably entertained in the immediate wake of September 2001, terrorism has remained a globally minor threat when compared with other causes of death. The tragic loss of 7,142 lives, globally, to terrorism in 2021,[21] for example, is hugely over-shadowed by the 25,000 lives lost to hunger every day.[22]

Partial strategic counter-terrorist success might also be judged to have been achieved if terrorists' narratives are prevented from gaining the day, and if terrorists' strategic goals are blocked and their own victory therefore prevented. Al-Qaida's central strategic objectives were to overturn what they considered to be apostate Muslim regimes, to expel the USA and US influence from Muslim countries, and to bring about the renaissance of a particular kind of Islam. None of these has been achieved, and determined US policy has helped ensure this fact.[23]

In terms of narratives, the case seems more complicated. There were some US successes. The framing of the post-9/11 effort as a war gained considerable momentum; even post-9/11 France (not always the most enthusiastic of American allies) very much endorsed the terminology of the struggle against terrorists being that of a war.[24] Moreover, in addition to their central strategic goals eluding them, al-Qaida failed to dominate the narrative; in 2013, 57 per cent of Muslims surveyed in a major global attitudes project held a negative view of al-Qaida, and only 13 per cent a positive one.[25]

Much of this involves credibility and—for all of the flaws in their approach after 2001—Western states have repeatedly tried to ground their counter-terrorism on a strongly plausible analytical foundation. Credibility, for example, involves accurately identifying the motivations of those practising terrorism; misdiagnosing the causes behind terrorists' behaviour is likely to make effective counter-terrorism elusive.[26] Accordingly, in 2006, Eliza Manningham-Buller highlighted both the Security Service's successes in stopping terrorist plots against the UK, and also the importance in that work of understanding terrorist adversaries:

There has been much speculation about what motivates young men and women to carry out acts of terrorism in the UK. My Service needs to understand the motivations behind terrorism to succeed in countering it, as far as that is possible. Al-Qaida has developed an ideology which claims that Islam is under attack, and needs to be defended. This is a powerful narrative that weaves together conflicts from across the globe, presenting the West's response to varied and complex issues, from long-standing disputes such as Israel/Palestine and Kashmir to more recent events as evidence of an across-the-board determination to undermine and humiliate Islam worldwide.[27]

There have, however, been clear weaknesses in counter-terrorist narratives since 2001. The damage done to US credibility through the flawed justifications for the Iraq War represents one example. Avoidable human rights transgressions embody another. I'll come back to some details of the latter, in relation to specific tactical-operational methods such as Unmanned Aerial Vehicles (UAVs), or the effects of mistreating prisoners. But the broad point can be adumbrated here, that some aspects of post-9/11 tactical work (whatever their other effects) undermined the narrative argument offered by the US and its allies in relation to counter-terrorism.

The record here is therefore not straightforward. Both al-Qaida and the post-9/11 US authorities sought to present the world as divided between good and evil, and as experiencing a conflict between the moral and the immoral; neither side proved as persuasive as they would have liked in this endeavour.[28] Jihadists such as Osama bin Laden failed to gather sufficient momentum within the Muslim world to achieve their own victory. But the USA and its allies undermined their own narrative through some avoidable errors of judgement, transgressing some of those rules of appropriate behaviour which characterize liberal democracies in their treatment of their own and other citizens.[29]

At the tactical level, there was again much that was complicated in the counter-terrorist record of the War on Terror. Operationally, there were many successes. A large number of al-Qaida actors were captured or killed; al-Qaida's base of operations in Afghanistan was removed during the US engagement there; jihadists' capacity to move money was challenged and undermined. The 2 May 2011 US killing of al-Qaida leader and icon Osama bin Laden was a high-profile tactical-operational success.[30] Twenty years after 9/11, the capacity of al-Qaida to maintain a residual terrorist threat remained; but those two decades had also provided clear evidence of the limitations on what they could achieve in practice, and where.[31] These tactical-operational successes were markedly visible in the early phase of action in, for example, Afghanistan. But they have continued. In December

2020, the UN Analytical Support and Sanctions Monitoring Team noted that al-Qaida had recently 'endured a period of high leadership attrition', with multiple deaths across numerous countries.[32]

In terms of US allies, the UK's tactical-operational experience is also worth considering. Between 9/11 and 2007 the UK faced fifteen major terrorist plots, of which twelve were thwarted by the authorities.[33] Between 2013 and 2018 the UK authorities claimed to have prevented twenty-five jihadist terrorist plots, and during 2017–18 four right-wing terrorist plots were also claimed to have been foiled.[34] Successes such as Operation Crevice in 2004 represent significant tactical-operational victories: here, MI5 and the London Metropolitan Police prevented, through coordinated surveillance, an al-Qaida-supported plot to carry out bomb attacks in Britain; five life sentences for the plotters followed in the UK, with one of the terrorists (a Canadian) being sentenced in Ottawa to more than ten years' imprisonment.[35] Operation Rhyme in the same year resulted in the prevention of further jihadist terrorist attacks, intended to cause mass casualties.[36]

Despite its more negative aspects, the early phase of the War on Terror had generated some gains in terms of security and in terms of damage done to al-Qaida.[37] Some of these tactical-operational successes have worked partly because terrorist organizations have vulnerabilities in terms of their communication processes: when terrorist leaders communicate with their colleagues in order to direct and control the latter's violent or financial activity, the communications make their organizations more vulnerable to state observation, penetration, interception, and capture. Here, the normality and mundanity of terrorists—the fact that they face managerial challenges similar to those experienced by other organizations—makes them less threatening and more vulnerable.[38]

Sometimes, the state's own culture can make counter-terrorism more difficult than necessary. The strong tendency to assess success through metrics-based analysis, for example, has been an enduring problem in US counter-terrorism. It is not that there is nothing that can be calibrated through metrics-based measurement; it is rather than many aspects of counter-terrorist success are hampered by an over-reliance on this approach. In the case of the CIA, for example, an exaggerated interest in rewarding what can easily be counted (the number of spies recruited, the number of reports produced) has steered behaviour away from some of the more subtle issues central to success (such as shrewd judgement about the actual value of a spy's information, experience-based intuition about what is crucial in a report, or liaison with other agencies in pursuit of counter-terrorism).[39]

Another criticism of tactical-operational counter-terrorism in these years might be that the enormous expenditure allocated to it has not been based on appropriate consideration of what these vast sums actually secure, and how proportional these are to the threat that is genuinely faced. Nearing the tenth anniversary of 9/11, US federal spending on domestic homeland security had already grown by over $1 trillion during that decade.[40] Did the gains justify the costs? Put another way, had this counter-terrorism money been spent on other things (such as health care or protection against natural disasters), would more lives have been saved? The terrorist threat to American lives was certainly minimal (even after 9/11), compared with other threats such as accidents or non-terrorist killings.[41] So the vastly increased spending after the 2001 atrocity seems arguably inefficient, and far from cost-effective.[42] Indeed, many of the measures introduced seem not to have had a significant effect on the terrorist threat at all.[43] It could perhaps be argued that the terrorist challenge to America has been low because of such huge counter-terrorist expenditure; but the vast gulf between daily danger and daily expense does seem significant. The statistical likelihood of a US citizen being killed by terrorism is remarkably low, and much lower than the threat they face from other causes of harm (such as accidents, diseases, natural disasters, and non-terrorist murders);[44] one 2016 study found that 'the yearly chance an American will be killed by a terrorist within the country is about one in 4 million under present conditions'; it also pointed out that, 'At present rates, a passenger's chance of boarding an airliner in the United States that is subsequently attacked by terrorists is something like one in 22 million.'[45]

It should also be stressed that there had been much higher levels of terrorism in the US during the 1970s than there were in the years after 9/11.[46] Yet between the fiscal years of 2001 and 2004, United States federal defence and security spending grew by more than 50 per cent, from $354 billion to $547 billion; the Afghanistan and Iraq Wars were important here, but so too was increased spending on homeland security.[47]

Intelligence represents another crucial area of tactical-operational work,[48] and the picture here has again been a mixed one. High-grade and well-deployed intelligence can provide important information about the strength and dynamics of terrorist opponents; it can help to prevent attacks; it can aid arrests and prosecutions; and the very process of gaining it through informers can also undermine terrorist organizations by creating doubt among them regarding their comrades, and by generating mistrust and suspicion among terrorists regarding one another.[49] The US National Security Agency (NSA) gathers millions of intelligence data daily.[50] Equally, substantial and enduring terrorist groups have sometimes developed their own counter-intelligence capacity,[51] again reflecting the significance of this tactical-operational field of activity.

Some modes of tactical operation have carried particular ambiguities with them and require specific reflection. Scholarly opinion varies regarding the efficacy of targeted killings, and assessing this is as difficult as it is controversial, not least because of the contested numbers and identities of those actually killed in attacks. There are those who have argued that their disruptive effects (secured through the removal of people with rare skills and knowledge) make targeted killing a tactically valuable approach within the countering of terrorist organizations. Others suggest that the tactic sustains or even intensifies the terrorist resistance that it is intended to uproot. Yet others argue that, in itself, the method makes little overall difference to levels of violence. In terms specifically of decapitation (the killing or capture of the key leader(s) of a terrorist group), opinions again diverge. Some argue that this tactic has significantly damaging effects for terrorist groups, whether in terms of hampering their operational capacity or in terms of leading to group defeat. Others argue that its efficacy is questionable, and its counter-productive consequences considerable.[52]

Whatever the complexities, decapitation has been a major operational tactic within many states' counter-terrorism strategy, including that of the twenty-first-century US. What seems fair to stress is that there is no certainty that leadership decapitation will lead to group defeat, nor that the tactic is likely to work on its own (independent of other counter-terrorist or political endeavours). Even those who write positively about decapitation's effects can acknowledge these points, as also the recognition that context must remain decisive, that incarcerating terrorist leaders might work as well as killing them, and that long-term as well as short-term effects must be considered when decisions are made about decapitation activity.[53] The repercussions from collateral damage have to be remembered when considering the positive, tactical-operational effects of these methods. Here, as so often, there is the danger that tactical efficacy generates problems at the strategic or partial strategic levels, in terms of hearts and minds, credibility, and narrative success.

Remotely Piloted Aircraft (RPA), UAVs, or drones represent a related case study in tactical-operational terms. There is little doubt regarding the technical qualities now inherent within drone manufacture and use, nor any question about their capacity to allow the identification and/or targeting of specific terrorist opponents or sites. Aircraft without a pilot on board, and controlled by someone on the ground, UAVs were used extensively in Afghanistan and Iraq, and have been a major weapon in the War on Terror more broadly. The lethal use of drones in counter-terrorism expanded greatly under President Barack Obama, who authorized hundreds of targeted killings. By 2014, the US military possessed more than 10,000 drones. The expansion of lethal UAV use reflected the tactical benefits that drones offered. But it also generated some profound anti-US opinion internationally, something that endured during Donald Trump's presidency (which saw, again, an expansion of drone deployment).[54] It is easy to see the appeal of RPAs for counter-terrorists: they reduce the risk to the attacking force, and can do much in terms both of reconnaissance and of targeted killing. There is an unquestionable tactical efficacy to drones: they can kill enemy targets without combat risk to the person deploying them. During the War on Terror, drones have therefore enabled the US and its allies to attack some key enemies (such as al-Qaida's military chief Mohammed Atef in 2001) and to degrade terrorist opponent groups significantly.[55]

Whether such strikes reduce the violence of the targeted groups is more contested.[56] As so often, there may here be a tactical-operational efficacy without there necessarily being a significant strategic or partial strategic benefit. Indeed, the possibility that drone attacks stimulate terrorist recruitment may even mean that tactical effectiveness and strategic self-damage co-exist for those using UAVs. There is certainly some compelling evidence that drone attacks by the West (especially those that involve civilian casualties) have generated considerable local hostility, recruitment to terrorist groups, and acts of terroristic violence against Western targets and allies.[57] As a result, there have been arguments made that drone attacks have ultimately proved a counter-productive part of Western counter-terrorism, especially because of the civilian deaths that have been caused by them.[58] It is also true, however, that local responses to drone use, even in Pakistan, are more varied than is sometimes assumed: there has indeed been much controversy and outrage, but there is also some support even here for drone use, as well as much ignorance of the technology's deployment.[59] In tune with the theme of terrorists and counter-terrorists mutually shaping and echoing each other's behaviour, it should also be noted that there exists the possibility of an increasing use of drones by terrorist actors themselves.[60]

A major 2014 Report from a Commission chaired by David Omand observed that drones 'now represent an increasingly important potential for the modern military', and recommended that the UK should aim for a situation in which their use 'is viewed as an integral, essential, and normal component of UK airpower'. Echoing concerns about the ongoing practical use of the technology, however, the Report also argued that, 'In situations where UK forces are embedded with US or other forces, the UK government should do more by way of reassurance to explain the safeguards which are in place to ensure that embedded personnel remain compliant with international humanitarian law.' Centrally, the injunction was to deploy drones (as other technologies) in accordance with proper principles of conflict: 'the ethical acceptability' of drone use, 'like that of other weapon systems, is contextually dependent upon meeting the legal principles of distinction and proportionality'.[61] This analysis helpfully avoided exaggeration of what this latest technology changed about the ethics and practice of counter-terrorist conflict; and it reflected the vital issues of restraint, proportionality, and legally responsible behaviour by states.

Relatedly, the tactical-publicity aspect of the War on Terror has been ambiguous. There have been major coups. The early, striking successes in Afghanistan in 2001 and in Iraq in 2003, or the killing of Osama bin Laden in 2011, all provide sharp-edged examples. But, just as there is a paradox to terrorist publicity (gaining high-level attention for acts that repel the majority of observers),[62] so too there has been a paradox to the high visibility of much of what has been done during the War on Terror. In particular, a major problem for terrorists and counter-terrorists alike since 2001 has been the damage done to their respective causes by their own violence against civilians.[63]

Damaging publicity also emerged for the US from its treatment of those detained in Guantanamo Bay and in Iraq's Abu Ghraib prison.[64] Some of this was disgraceful (though surely predictable in a setting such as war-time Iraq?), and the publicity understandably given to images of prisoners being abused has had a lasting and undermining effect on the US and its reputation.

Torture specifically requires some discussion here. There is no doubt that the US and some of its allies on occasions deployed torture during the War on Terror.[65] The 9/11 attack and the fear of further terrorism prompted the

USA to engage in such activity with terrorism suspects.[66] In the aftermath of September 11, President Bush's regime had claimed that the unprecedented exceptionality of the threat and of the enemy that they now faced was such as to allow the USA to be exempt from certain norms of conduct. In the 1970s, US regimes had tended to see terrorism as a threat best interpreted and dealt with as a crime and therefore through law enforcement mechanisms.[67] After 9/11, al-Qaida were judged to be political enemy combatants, but unlawful ones. This approach did not go uncontested within or beyond the US government,[68] and nor did the implications of such a view (such as the use of torture against terrorist suspects). But it did lead to some brutal activity. Sometimes, detainees were transferred through extraordinary rendition: the transporting of prisoners from one country to another without formal extradition processes being used, to countries with dubious human rights records, in order to facilitate torture-based interrogation.[69]

On occasions, some useful information may perhaps have been acquired. But the overall effect of this practice was negative in at least two main ways. First, experts from many different backgrounds and perspectives have judged interrogational torture not to be the best way of producing the accurate, new, timely, significant, reliable, actionable intelligence that is required in successful, modern-day counter-terrorism.[70] Indeed, second, the experience of mistreatment in jails is more likely to generate terrorism than to reduce it, and to create future hostility and attacks against the state in question.[71] Transgressing those laws which distinguish liberal democracies from more ruthless regimes and actors has long been known to undermine states in their credibility in counter-terrorism.[72] But this is what happened on far too many occasions during the War on Terror. It should be recognized, of course, that of the thousands of post-9/11 US interrogations (in Iraq, Afghanistan, Guantanamo Bay, and elsewhere), most did not involve the use of torture.[73] It should also be acknowledged that, after much sharply polarized debate, the argument within the US for normalizing the expansion and utilization of torture in counter-terrorism was eventually lost.[74]

But there remains little doubt about the self-undermining aspect of the West's mistreatment of some (especially Muslim) prisoners during especially the early phase of the War on Terror.[75] In this and related ways, US-led counter-terrorism in recent decades can be judged to have undermined important aspects of democracy, human rights, liberty, and the rule of law.[76] While such transgressions still leave Western states some distance away from the kinds of practice common in authoritarian regimes, they remain tactically damaging in terms of publicity, especially given the importance of revenge as a motivation for ongoing terrorism.[77] This should not be the main reason for opposing such mistreatment; but it remains an important one.

Damaging publicity was also generated through tactics made famous by people like the NSA analyst and private contractor Edward Snowden. In 2013, Snowden acquired and leaked large amounts of classified information to journalists regarding the digital intelligence work of the US and UK. Stories about the mass gathering of electronic data were duly published, and independent inquiries were launched in both countries. In 2015, a US appeals court ruled that the PATRIOT Act[78] did not authorize the gathering of bulk metadata; in the same year, the US Congress acted to limit mass surveillance. Similarly, the 2016 UK Investigatory Powers Act embodied an attempt to regulate appropriately what is done in relation to surveillance by the state. Commenting on Snowden's sharing of NSA activities and data with the press, former FBI Director James Comey observed that, 'One obvious result of this theft was that it dealt a devastating blow to our country's ability to collect intelligence. Another result was that, in the year after his disclosures, bad actors across the world began moving their communications to devices and channels that were protected by strong encryption, thwarting government surveillance, including the kind of court-authorized electronic surveillance the FBI did. We watched as terrorist networks we long had been monitoring slowly went dark, which is a scary thing.'[79]

Snowden was charged by the US under the Espionage Act. By the time of his leaks, however, he had left the US, and at time of writing he remains in Russia, where he was granted permanent residency and eventually Russian citizenship. The fact that a 29-year-old junior analyst could gather and release such vast amounts of secret material hardly suggests the most oppressive system or culture of surveillance to have obtained in the US in 2013. Moreover, there have been serious concerns about the negative effects of Snowden's actions. The material he took contained information about personnel and also about the methods of surveillance used to protect the population from terrorist attack.[80] On the other hand, some aspects of the issues dramatically raised by Snowden's actions (especially regarding important questions of proportionality, transparency, and efficacy) were unquestionably significant.[81]

In terms of the tactical work of undermining opponents, much was clearly done during the War on Terror. This includes the attacks and operations already mentioned involving the degradation of al-Qaida, ISIS, and other terrorist groups, but also the frequent successes that have been achieved in undermining terrorist organizations' finances.[82]

In terms of population control, in addition to the clear success of the US and Western allies in maintaining their own effective domestic legitimacy and power, there were considerable tactical successes in the occupation phases of the wars in Afghanistan and Iraq, though—as noted—later developments called some of this into question. Regarding the final tactical category, organizational strengthening, the picture must again be seen as mixed. The US's 9/11 Commission Report judged there to have been major failures, each of which had contributed to facilitating the possibility of the 2001 attacks: 'We believe the 9/11 attacks revealed four kinds of failures: in imagination, policy, capabilities, and management.' Under the fourth of these headings, the issue of stove-piping among intelligence and other agencies was highlighted; too often, different agencies worked in isolation from one another, rather than as a team.[83] There can be little doubt about the damage done to pre-9/11 US intelligence capacity by this lack of organizational coordination.[84] After 9/11, there was reorganization intended to create greater counter-terrorist coordination and resourcing, and some real change was indeed effected here. The US Department of Homeland Security (DHS) was established following the 9/11 attack, with the goal of coordinating homeland security efforts. In September 2019, the DHS Acting Secretary claimed, vis-à-vis foreign terrorist organizations, that 'we have made great progress in our ability to detect, prevent, protect against, and mitigate the threats that these groups pose'. Part of this, he claimed, involved the fact that, 'We have increased the sharing of information about terrorist threats between the Federal Government and state, local, tribal, and territorial entities, as well as private sector partners.'[85]

Indeed, there were repeated attempts at developing strong partnerships beyond the state itself. In the UK, the CONTEST strategy explicitly committed itself to 'an approach that unites the public and private sectors, communities, citizens, and overseas partners'.[86] Collaboration and cooperation between different wings of the state, and between different states, are crucial to the achievement of tactically successful counter-terrorism, and it is well-established that counter-terrorist failures have often emerged or been deepened by a lack of such coordination.[87] Where inter-agency cooperation is achieved (within and also between states), success tends to be far more likely;[88] where there is divisive compartmentalization, counter-terrorism is made more difficult.[89]

In the US, this kind of organizational strengthening and coordination after 2001 has indeed had some success, but with marked limitations too.[90] Pre-9/11 US counter-terrorist failings had undoubtedly played their part in facilitating the 2001 atrocity. There had been indications during the summer of 2001 that a major terrorist attack was looming, but there was a failure to coordinate effectively and to act decisively in ways that might have prevented the 9/11 assault. Fragmentation problems within the world of US counter-terrorism, together with a failure by the US government to respond to al-Qaida as a high-level threat, played their part here. And if organizational dysfunctionality made it more difficult for the US authorities to prevent 9/11, then it is also true that (despite some subsequent progress being made) crucial organizational problems endured during the post-9/11 period, particularly in terms of a lack of effective integration and efficient coordination and cooperation between different wings of the enduringly fragmented, disorganized, and ill-designed US intelligence community.[91] The complexity of the state itself, with its multiple elements and actors, can make things more challenging here;[92] and the bureaucracy of US security infrastructure is notoriously cumbersome, disjointed, and fragmented[93] (so much so that this problem has long been identified even in popular literature, as has also been the case for the UK).[94]

Within the UK, the 2003 establishment of the Joint Terrorism Analysis Centre (JTAC) represented a major effort to coordinate multi-agency approaches to terrorism, and to provide collaboratively produced assessments and threat level appraisals for the government. International collaboration has represented a complementary aspect of organizational strengthening, and the Five Eyes network (an intelligence network comprising the US, UK, Canada, New Zealand, and Australia) has produced repeated tactical benefits. Other international alliances have offered profound challenges as well as rewards (a partner such as Pakistan, for example, being simultaneously ally and adversary for much of the post-2001 period).[95]

This detailed consideration of the overall strategic, partial strategic, and tactical outcomes of the War on Terror therefore shows them to have been immensely complex, and immune to simply condemnatory or celebratory response.

What of inherent rewards? There was emotional catharsis and political benefit derived from major counter-terrorist operations such as the 2011 killing of Osama bin Laden. Though none will prove as famous as this, other decapitation attacks against al-Qaida and other enemies unquestionably offered some inherent rewards for those engaged in authorizing or operationalizing them. Domestic political support can be garnered by politicians who engage in aggressive counter-terrorism policies;[96] President Bush's immediate response to 9/11, trenchant and popularly resonant in the USA as it was, led his approval ratings to rise to over 90 per cent.[97]

The increasing use of drones or UAVs (beyond the attractions of their lethal potential) can lead to economic benefit to certain parties—from their development, production, and sale.[98] On a different level, drone pilots themselves can be subject to a mixture of inherent rewards (excitement, pride, revenge) and moral concern and anxiety over the collateral damage inflicted.[99] I've mentioned some of the financial gains people made from the Iraq War, but the pattern is a wider one than that. Indeed, economic benefits from post-9/11 counter-terrorism have been truly enormous for some: 'A 2011 study by the Pentagon found that during the ten years after 9/11, the Defense Department had given more than $400 billion to contractors who had previously been sanctioned in cases involving $1 million or more in fraud.'[100] Less dubiously, in pursuing technological means to increase post-9/11 security in the USA, considerable business opportunities were generated. The security technology market expanded vastly as the US government's technology spending increased. University campuses as well as technology laboratories saw a huge growth in innovation and business development, with enormous funding opportunities emerging as a result (in relation, for example, to research, development, and production of biometric sensors, innovative security software, surveillance technology, and drones).[101]

Emotional and psychological rewards could also be multifaceted. Those involved in counter-terrorism can exhibit a clear pride and enjoyment in the value and significance of their work, as well as a sense of the excitement and shared motivation that are sometimes involved in it.[102] In terms of the US military and counter-terrorism community, it is clear from firsthand and other evidence that there have existed considerable benefits, including comradeship, shared loyalty, a sense of altruistic commitment to a righteous cause,[103] and the satisfaction of a desire for revenge against terrorist enemies.[104] Revenge could also motivate politicians, as with President Bush's comment soon after hearing of 9/11: 'We are going to find out who did this, and kick their ass.'[105]

Clearly, therefore, the post-9/11 War on Terror had both successes and limitations.[106] The evidence points to a very complex reality.

Full strategic victory eluded the US and its allies; the rhetoric of destroying the evil of terrorism, of driving it from the world, of eradicating jihadist terrorism, and of stopping, defeating, and eliminating global terrorist groups had set the agenda too high.[107] For, given these stated ambitions, a form of success can be claimed by terrorist organizations who survived and thereby thwarted these full strategic ambitions.

Some partial strategic success could legitimately be claimed by the US and its allies: in terms of sustaining the normality of societal life; in terms of the considerable degradation and damage that were done to al-Qaida, ISIS, and numerous other terrorist adversaries; and in terms of preventing terrorist enemies' victory and preventing the achievement of their central goals (certainly with regard to al-Qaida and ISIS;[108] though far less so, at time of writing, with the revivified Taliban in Afghanistan). Determining the global agenda and narrative was far more ambiguously pursued, and the long-term effects of what many will see as ultimate terrorist victory in Afghanistan may be pernicious in this regard. Moreover, the secondary goals of building stable, region-transforming, benign democracies in Afghanistan and Iraq remained disappointingly elusive. In terms of counter-terrorism, the Iraq experience was largely negative, for the reasons I've rehearsed above (removing Saddam Hussein did not displace an al-Qaida-supporting regime which possessed WMD, the war stimulated much terrorism, and damage was done to US credibility).

The tactical picture which emerges is complex: there were very many operational successes, and these deserve respect and appreciation in terms of the lives that they have saved; interim concessions could be detected in the way that even the Taliban allowed some unchallenged departures during the 2021 withdrawal from Afghanistan; publicity could be positive (especially with high-value successes such as the killing of Osama bin Laden and Ayman al-Zawahiri) but was often negative because of collateral damage, mistreatment of prisoners, and false claims; terrorist opponents were often undermined, and domestic populations largely remained supportive of the normality-sustaining approach of Western governments. There was some coordinating and strengthening of the organizational infrastructure and resilience of the state in its counter-terrorist work, but it is also true that stove-pipe problems remain across many Western state organizations. Inherent rewards were many (politically, financially, professionally, and in terms of revenge), but large numbers of people paid a high cost, including the many thousands of US and allied forces killed and maimed, let alone the damage done to so many others in Afghanistan, Iraq, Syria, and beyond.

II

In all of this, some major themes have emerged. One is the tension between tactical and strategic dynamics. Drones have offered extraordinary and unprecedented tactical opportunities, and have been extremely attractive weapons to the authorities as a result; but they have also brought with them long-term risks in relation to crucial aspects of counter-terrorist success at strategic or partial strategic level, such as the sympathy and support of vital populations in conflict and other politically relevant settings. Military capacity more broadly—as evident in the extremely impressive early victories in Afghanistan in 2001 and Iraq in 2003—has again attracted energy and become central to counter-terrorist thinking after 9/11; but there has at times been an exaggeration here of what military means can achieve.[109] Indeed, the early successes in those two war-zones (Afghanistan and Iraq) reinforced a hubristic lack of concern with the longer-term political and cultural work necessary for strategic success. Ironically here, counter-terrorism and terrorism again echo one another. The 9/11 atrocity was a tactically remarkable success for terrorism, which did very little to advance al-Qaida's central strategic goals; indeed, in terms of subsequent US involvement in Muslim countries, the attack proved strikingly counter-productive. For both sides in the 9/11 wars, tactical potential and success have distracted attention from what is centrally and strategically at stake. Many lives have been lost as a consequence.

A second major theme which emerges from this story is that successful counter-terrorism needs to take account of the wider politics, society, culture, and history which both explain why terrorist violence has emerged (that is, the deeper problems of which terrorism is a symptom), and also prepare states for effectively dealing with it. As practitioners have rightly observed, counter-terrorism needs to be integrated into wider political patterns and to avoid a narrow or discrete approach.[110] Terrorism is not in fact the main phenomenon to be considered when pursuing successful counter-terrorism. Yes, terrorist organizations, operations, intentions, activists, and plans need to be countered. But the context of rival political causes and their long-complex pasts must be apprehended as well, if one is genuinely to understand terrorism and counter-terrorism.[111]

If (as after 9/11) there is a misdiagnosis of the causes and motivations behind the terrorism to be combated; if there is an amnesia about the known knowns of effective counter-terrorism inherited from its long past; if there is an insouciance about the culture and history of countries being invaded—if all this, then counter-terrorism will become avoidably difficult. And that is precisely what occurred.

Third, post-9/11 counter-terrorism was far too short-termist in approach. Not only were the histories of terrorism and counter-terrorism (and of Afghanistan, America, and Iraq) substantially ignored in a disgraceful fashion,[112] but the long futures implied by long pasts were also too often ignored. This was evident in the lack of preparation for post-war Iraq. But it was also clear in the apparent disregard shown for the counter-productive side effects of tactics such as drone attacks abroad. The idea of recasting the Middle East through naive invasion ignored the profound and well-established complexities involved in that region's inheritances, allegiances, and contingencies.[113]

Some US rhetoric did suggest appreciation of long-termism. President Bush in October 2001 said that the US would require 'patience and understanding that it will take time to achieve our goals';[114] in December 2001, he suggested that, 'Preventing mass terror will be the responsibilities of Presidents far into the future';[115] and in January 2002, he pointed out that, 'Our War on Terror is well begun, but it is only begun. This campaign may not be finished on our watch.'[116] The continuities of approach among Presidents Bush, Obama, and Trump should not surprise us in this context.[117] But a more profoundly long-termist approach was required, one which more seriously considered the implications of thinking historically about US, UK, and other Western experience, and about the long futures for which foundations were being built. In all of this there was too often a lamentable inadequacy of approach much of the time, and that limited the likelihood of success.

III

There is a final twist to this story. Honest consideration of two decades of post-9/11 counter-terrorism led by the US necessitates recognition both of how deeply embedded terrorism and counter-terrorism have been in long-term US experience, and also of the extensive terrorism which threatens Americans not from jihadist violence (the overwhelming focus of the War on Terror) but from right-wing American extremists.

Domestic, far-right terrorism became conspicuous as President Trump was leaving office in 2021, a development not unrelated to the ambiguity that he had frequently evinced regarding the issues on which such actors base their arguments. Here was a president who had shown little interest in addressing the root causes behind jihadist terrorism, yet who seemed sympathetic to the concerns of those on the US nationalist far-right, some of whom clearly threaten and/or use terrorist violence on the basis of those concerns and grievances.[118]

The historical roots of American terrorism are far longer and sturdier than much public discussion seems to suggest. The violent far-right in the US has a long tradition (the Ku Klux Klan embodying a white supremacist example, as has rightly been noted in recent US strategy pronouncements).[119] And, as historian Michael Fellman pointed out, while many Americans have preferred to see terrorism as something exceptional and external, terrorism in the US has in reality been deep-rooted, diverse, often domestic in origin, and regrettably pervasive.[120]

Violent, right-wing extremism in the US is fissiparous rather than monolithic,[121] and this kind of sub-culture is vitally important for counter-terrorism for two main reasons. First, deeper recognition of how far into its past the US had had to deal with terrorism of numerous kinds would have encouraged a less exaggerated and panic-driven response to 9/11 than occurred (again, the problem of ahistorical counter-terrorist thinking is all too evident here). Much of what went wrong in the War on Terror (the abandonment of appropriate legal restraints in treating suspects; the over-militarization of counter-terrorist tactics; the unrealistic ambition to extirpate international terrorism; the misuse of intelligence) was made more likely because of a failure to recognize the continuities of experience across the supposedly stark 9/11 fault-line. What constituted effective counter-terrorism did not, in fact, change dramatically when the planes flew into the towers on that terrible Tuesday. The mistaken assumption that everything was now altered, and that a new terrorist era had been inaugurated, was one which facilitated many of the most egregious mistakes of ensuing decades; and amnesia regarding the country's long experience of terrorism made it easier to misdiagnose, exaggerate, over-react, and misrespond to the post-9/11 threat.[122]

Second, disproportionate attention became focused after 9/11 on what was a comparatively limited threat to most Americans (jihadism), and too little notice was taken of a more significant terrorist danger to life and limb, in the form of right-wing violence. It is not that US authorities have simply ignored right-wing terrorism. In 2019, the Acting Secretary of the DHS stressed 'the dangers posed by domestic terrorists, including racially- and ethnically-motivated violent extremists, particularly white supremacist violent extremists'.[123] But it is the case that far more attention was focused after 9/11 on jihadism, and so the eye-catching events of early 2021 in Washington, DC, represent a painful epilogue to our consideration of post-9/11 America's attempt to counter terrorist violence.

The 6 January 2021 attack on the US Capitol by supporters of the outgoing President Trump was striking for its scale, for its intention of overturning a presidential election, and (thankfully) for its essential amateurism. A large crowd of Trump enthusiasts had gathered in Washington; some of them attacked the Capitol building with the aim of preventing the certification of Joe Biden's victory in the recent presidential election; the siege lasted over four hours, and five people lost their lives in the violence. Terroristic attacks on the Capitol had occurred previously in American history, and one important aspect of this disturbing 2021 assault was the extent to which it reflected the longer-term and extensive far-right culture of terroristic violence in the US.[124] As always with significant terrorism, the violence is a ghastly symptom of what is more profoundly occurring. The sense of economic and cultural marginalization, displacement, and replacement that is felt by many who are supportive of such ventures is complemented by various other elements: a tradition of defending liberties that are perceived as being under threat from an over-intrusive state; a religious motivation for some; racial antagonisms and fears for others; long-rooted American militia cultures; a profound sense of lost status and rights for white Americans; and an instinctive hostility to the Federal government.[125] 'Today's domestic terrorists espouse a range of violent ideological motivations, including racial or ethnic bigotry and hatred as well as anti-government or anti-authority sentiment. They also take on a variety of forms, from lone actors and small groups of informally aligned individuals, to networks exhorting and targeting violence toward specific communities, to violent self-proclaimed "militias".'[126]

The 2021 Capitol assault reflects a terrorist problem that is unlikely to dissipate speedily. By 2022, the growth of right-wing militia activity in the US had become very striking. The authorities have had some success in thwarting planned violence (as with the intercepted plot against Michigan governor Gretchen Whitmer in 2020). On other occasions, however, the state was unable to prevent attacks, and the majority of post-2008 extremism-related deaths in the US have been generated by the right.[127] Attacks included the mass shooting on 14 May 2022 in Buffalo, NY, in which Payton Gendron killed ten people in an assault deliberately targeting Black people; FBI Director Christopher Wray described the atrocity as 'a targeted attack, a hate crime, and an act of racially motivated violent extremism'.[128]

Some observers have been surprised by the vehemence and extent of right-wing terrorism in the United States. Again, however, this involves regrettable amnesia. For all of the explicable attention focused on jihadist enemies after the 9/11 attack, the depth and extent of right-leaning violence has been one of the numerous strains of long-rooted terrorism within the US. *The Turner Diaries* (the 1978 novel by William Luther Pierce (1933–2002), published under the pseudonym Andrew Macdonald)[129] depicts a white-racial revolution in the US, one that leads to nuclear war and to the eradication of most non-white people from the earth. The book has become celebrated and influential among some more recent right-wing terrorists, including Oklahoma bomber Timothy McVeigh (whose own 1995 attack was strongly pre-echoed in one of the terrorist incidents fictionally detailed in *The Turner Diaries*).[130] Indeed, Pierce's book's themes evoke much that has been involved in an enduring right-wing US terrorist culture, and they reflect the historically long roots of such thinking and behaviour: the revolutionary 'Organization', with its inner-circle group, the 'Order', at war with the liberal, multi-racial US 'System'; the theme of requiring vanguardist, violent, brutal action in order to stimulate necessary political change and to jolt society into transformative action ('After all these years of talking—and nothing but talking—we have finally taken our first action. We are at war with the System, and it is no longer a war of words'; 'The future of our race depends upon the outcome of this war'); activist exhilaration in the struggle and its activities; a profoundly anti-Jewish, anti-Black, anti-Hispanic politics; an intense commitment to maintaining one's right to possess guns and ammunition; deep antagonism towards a multi-racial, oppressive, corruptly bureaucratic, intrusive state, and one that is interpreted as being hostile to the interests of an increasingly displaced white people (the novel states 'Pasadena used to be mostly White a few years ago, but it has become substantially Black now'); and a revolution that can change all of this.[131] Embedded within a long-rooted right-leaning culture within the US, Pierce's repellent, deeply offensive book both drew on a tradition of racist writings and itself directly and lastingly inspired terrorist actors.[132]

Right-wing terrorism and white supremacism are not synonymous in the US, and much non-jihadist terrorism in the American past has been from sources other than the right.[133] But it remains a gruesome twist at the end of two decades of trying to rid the world of terrorism that an alarmingly prominent terrorist threat to the US comes from such a long-rooted, domestic source, and one which the post-9/11 War on Terror did virtually nothing to erode.

Map 2 Northern Ireland

PART
II

Northern Ireland

4

War and Peace 1969–2005

I

The late twentieth- and early twenty-first-century Northern Ireland Troubles had extraordinarily long historical roots. In relation to the modern-day campaign of the Provisional Irish Republican Army (PIRA), one of the organization's US-based gun-runners informed me that, 'The Brits—they're the problem, and will be. They have been since 1169 [the year of the Anglo-Norman invasion of Ireland], and will be until such time as they leave.'[1] This very long past points to numerous crucial realities as we now consider the efficacy of recent UK counter-terrorism in the North of Ireland. It clarifies that the most vital issues at stake are not the terrorist symptoms (appalling though these have unquestionably been) emerging from competing nationalisms, but rather the politics and relationships associated with those rivalrous nationalisms, and the major questions over which they have been in conflict (state legitimacy and power; religiously fuelled national identities; the extent, nature, and legacy of colonialism in Ireland).[2] Historically long roots also hint at the need to be realistic about what can be expected when trying to limit or end the terrorist violence emerging from such powerfully enduring enmities. Very long timelines are likely to be involved when trying to address such conflict. With historical long-rootedness comes also a complexity of nuance which challenges more rigidly Procrustean attempts at understanding. The asymmetries between adversaries, the mosaics of intra-communal dynamics, the fluid nature of political organizations and traditions as they have changed over long pasts—all of this necessitates careful, historically minded analysis. So too it is worth noting in regard to Irish nationalist terrorism that, despite attention-grabbing violence from some of its adherents, nationalism in Ireland has more commonly been expressed through constitutional, non-violent politics than it has through terrorism.[3]

This section of the book focuses on the Northern Ireland Troubles which arose in the 1960s and which persisted into the twenty-first century. But, even in terms of the UK's experience of dealing with Irish-based terrorisms, there was a long pre-Troubles record. Indeed, attempts by English authorities to deal with Irish political violence preceded the creation of the UK itself, as in the case of the 1798 rebellion.[4]

In relation to nineteenth-century Irish republicanism, UK counter-terrorism involved dealing with attempted insurrection in Ireland, with bombings in England, with transnational networks of support, with the challenges presented by militant prisoners, and with the generation by rebels of legacy-building episodes and iconic figures.[5] It is also true that, in its struggle against those who sought an independent Irish republic through violent methods, the UK had long utilized spies and informers within Irish republican movements, and indeed had done so to considerable effect. This was true of Fenianism, and one effect of Fenian violence in England was to prompt the growth of counter-terrorist police operations there; but the tactic preceded even that.[6]

Early twentieth-century Irish republicans practised various forms of violence (as did their loyalist opponents and the forces of the state), most notably during the Irish revolution of 1916–23.[7] State reactions here could sometimes be clumsy, as was famously the case with the heavy-handedly counter-productive British response to the 1916 republican Easter Rising.[8] In that instance, state action contributed to the intensification of republican hostility to British rule in Ireland, and increased sympathy for the rebels whom it was intended to undermine.[9] Sharp-eyed Irish republican militants recognized the utility of prompting the British into self-harming over-reaction and harshness.[10] Of the British forces in Ireland in 1921, one leading Irish Republican Army (IRA) man later claimed that, 'Their campaign of terror was defeating itself'.[11] In another IRA leader's words about the same revolutionary year, 'the British reprisals, instead of turning the people against us as the cause of their miseries, had thrown them strongly behind us'.[12]

The term civil war was used by some contemporaries regarding early twentieth-century Irish political division.[13] This usefully signals one important dynamic of what was to follow in the 1960s and beyond in Ulster, namely the role of intra-Irish divisions amid the violence. The 1920s partition of Ireland, whose legacies would be so central to the modern-day Troubles, centrally occurred because a significant and sufficiently concentrated body of people in Ireland wanted to remain in the UK, whereas the majority of people on the island considered British rule to be illegitimate.[14]

The partitioning of Ireland clearly involved more than just the countering of terrorism. But the attempt to limit and prevent non-state violence in Ireland did form a major part of the dynamics involved and, while I don't think the emergence of the later Troubles to have been inevitable,[15] the eventual failure of partition to assuage rival nationalist grievances or to prevent future violence does demonstrate the importance of taking a long-term view when assessing attempts to deal with terrorist activity. If many republican activists from the revolutionary period failed to get the Ireland that they sought,[16] then ultimately the UK state's attempt to deal with Irish terrorisms through the mechanism of partition does partly have to be assessed through the blood-stained lenses of the subsequent Northern Ireland Troubles.

Those Troubles involved the transition from a civil rights movement, through inter-communal polarization and somewhat heavy-handed state response, into a civil war which lasted for decades and which cost nearly 4,000 lives. Anti-state terrorists such as the PIRA (Provisional Irish Republican Army) and the Irish National Liberation Army (INLA) sought to defend their communities from loyalist attack, and to resist, subvert, and destroy a Northern Ireland polity which they considered both illegitimate and systemically unfair.[17] Pro-state terrorists such as the Ulster Defence Association (UDA) and Ulster Volunteer Force (UVF) sought to bring pressure to bear, both on their Irish nationalist neighbours and on the UK state, to prevent the ending of UK sovereignty over Northern Ireland.[18] There was also violence from state forces (whether the British Army or the Royal Ulster Constabulary (RUC)), in efforts to contain, combat, and thwart non-state terrorisms.[19] The decades-long conflict was largely brought to an end through a peace process which reached its apotheosis in the 1998 Belfast/Good Friday Agreement, but which also involved important subsequent amendments and developments in attempts to sustain peace in Northern Ireland. Important here were political efforts by Irish nationalists within and beyond Northern Ireland, by unionists in Ulster, by Irish and British political parties, by international actors, and by many people at different levels of societal, communal, and political engagement.[20]

These peace-making endeavours helped provide the context and shape for the diminution of the Northern Ireland Troubles. But one important part of the long-term story was a counter-terrorist one.

II

Did the UK state secure strategic victory in its counter-terrorist efforts in Northern Ireland, with the achievement of its central, primary goal or goals? Strategic victory here would involve effectively removing the threat of more than trivial terrorism from its territory and people, and thereby getting rid of (almost all) terrorist violence. Perhaps more realistically, it might see a particular (major) terrorist organization becoming effectively neutralized as a danger. It might also involve a near-comprehensive political resolution of the conflict from which terrorism had emerged as a violent symptom.

Some have indeed claimed that the UK authorities defeated the Provisional IRA, the organization responsible for more killings than any other in the conflict.[21] Former RUC officer William Matchett has claimed that, 'Winning the intelligence war beat the IRA'.[22] Coming from a very different political background, former PIRA volunteer Anthony McIntyre has observed that:

> The political objective of the Provisional IRA was to secure a British declaration of intent to withdraw. It failed. The objective of the British state was to force the Provisional IRA to accept—and subsequently respond with a new strategic logic—that it would not leave Ireland until a majority in the north consented to such a move. It succeeded.... The strategic logic of engaging with republicans from the British point of view was to establish a process that would be inclusive of republicans but would exclude republicanism. The extent of British success can be gauged by the ground conceded by republicanism. The consent principle[23] and by logical extension, partition, has been accepted; the RUC has been modernized; the northern parliament has been re-established. Sinn Fein stands poised to prove republicanism wrong—and demonstrate that the northern state can be reformed.[24]

On such readings, the UK state effectively defeated its main terrorist adversary (the anti-state PIRA). From a different but also resonant perspective, speaking in 2021, MI5 Director General Ken McCallum presented a positive view of what had been achieved in Northern Ireland, through a combination of peace-process politics and effective counter-terrorist tactics:

> There is much to celebrate: the rejectionist terrorist groups[25] are much smaller now, they hold no meaningful mandate from the communities they pretend to represent, and, while they remain determined to cause harm, they continue to be subject to skilled, effective, proportionate action by security authorities

on both sides of the border. . . . I am and I remain a long-term optimist on Northern Ireland. The 1998 Belfast Agreement and the long process which led up to it, stands as one of the finest public policy achievements of my lifetime. It has enabled a whole generation to grow up substantially free of the scarring which haunted previous generations. The holding of multiple identities— British, Irish, Northern Irish—is a living reality for many people, in a way it was not in my youth. Those are deep shifts, which make a return to Troubles-scale terrorism highly unlikely.[26]

On such readings, the combination of counter-terrorist and wider political work led to substantial peace in Northern Ireland through the large-scale removal of terrorism from it. Both the UK Conservative and Labour Parties during the Troubles sought to combine security, constitutional, economic, and social approaches to the conflict.[27] Might such an approach be seen as having generated counter-terrorist success at strategic level?

There has certainly been a remarkable record in recent years in terms of the reduction of terrorist violence.[28] The December 2021 Independent Reporting Commission (IRC)[29] contained some very positive news here. Terrorism-related deaths in Northern Ireland between 2011 and 2021, for example, totalled twenty-three. The Report also said, however, that terrorism had not disappeared: 'paramilitarism remains a clear and present danger'. The reasons behind this were judged to be varied:

> paramilitarism involves a complex landscape comprising different categories of people. These range from individuals and groups who use paramilitarism as a cloak for overt criminality, to others who get caught up in it for reasons to do with socio-economic disadvantage. Some of this is related to the lack of an alternative pathway, to poor educational and employment opportunities, challenges to do with drugs, as well as addiction and mental health, and peer pressure. There are also 'dormant' members who retain some form of affiliation with a group (perhaps willingly, or perhaps because there is not a means to exit the group), who may pay a membership fee, and whilst not presently active in the organisations could be asked to play a role in the future. We consider that there is a further category of those who remain involved in paramilitarism for political and identity reasons which reach back to the Troubles. There is also a group of people, including at leadership level in some of the groups, who appear to have a positive wish to see paramilitarism ended, but cite a range of reasons why they believe they cannot do so currently.[30]

Almost twenty-five years after the 1998 Belfast Agreement which allowed for most terrorist violence to end, therefore, paramilitary organizations still existed in both the republican and loyalist communities within a deeply polarized and self-segregating society.[31] At the time of writing, the UK Security Service's 'Current Northern Irish Related Terrorism in Northern Ireland threat level' was such that: 'The threat to Northern Ireland from Northern Ireland-related terrorism is SUBSTANTIAL'; 'SUBSTANTIAL means an attack is likely'.[32]

The UK military's own analysis suggested that the state's goals during the 1980s included 'the destruction of PIRA'.[33] It could be claimed that by 2005 (when the Provisionals declared their campaign to be finished) the UK had indeed secured a major aim in regard to this enemy. But it was also true that the persistence of continuing IRAs (smaller, but still possessing occasional lethality and also some connections to the Provisional past)[34] slightly qualified this strategic success.

III

It would be naive to think that major terrorist organizations fully depart the stage quickly, and this is certainly crucial to acknowledge in relation to Northern Ireland.[35] As former Police Service of Northern Ireland (PSNI) Chief Constable Hugh Orde has put it, 'Endgames in terrorism are messy.... As most endgames, the Northern Ireland settlement was a negotiated one, but the negotiations had to continue post-event if we were ever to underpin the [1998] Good Friday Agreement with a sustainable and lasting peace.'[36]

But the achievement is very striking nonetheless. In 2007, the IRA's political party Sinn Fein even committed to supporting and engaging with the PSNI, a dramatic shift for an anti-state Irish republican movement to make. A case could therefore be made that the UK state secured strategic success in Northern Ireland, achieved partly through the counter-terrorist tactics that I will detail shortly, but also through partnership, negotiation, concession, and the complex politics of the lengthy Northern Ireland peace process. Even if one doubts such an argument, a very strong case indeed could be made that the UK state's counter-terrorism did help it to secure partial strategic victory, one within which it reduced terrorist capacity, substantially managed to protect the security of its people, and ultimately maintained order and peace. There were many aspects to this, but four important points need to be recognized: what the UK state long thought and wanted; the extent to which this was achieved in the Northern Irish peace process, with eventual endorsement from the PIRA and their political party Sinn Fein; the degree to which this endorsement involved Irish republicans departing significantly from what they had so long stressed as essential for an end to the conflict; and the complex processes (including, but not purely involving, direct counter-terrorism) which led to this republican sea change.

Let me take these points in order.

It had long been argued by the UK that, if the PIRA were to eschew violence, then a different kind of political engagement and dialogue could become more possible: 'The reason for our animus towards Sinn Fein is their expressed support for terrorism.'[37] This had also been a point long made by the PIRA's constitutional nationalist rivals in the Social Democratic and Labour Party (SDLP), who maintained for years that non-violent means existed for the better pursuit of political change. In 1974, for example, that party had declared that, 'The Provisional IRA can achieve nothing by carrying on their campaign of violence but they can achieve almost anything they desire by knocking it off.'[38]

While the PIRA did sustain their campaign, however, the UK state sought to limit the levels of violence in Northern Ireland, with a view to creating the conditions within which a political resolution could be achieved. Much maligned for saying it, the British Home Secretary Reginald Maudling suggested (at a press conference in Belfast on 15 December 1971) that he anticipated that the PIRA would 'not be defeated, not completely eliminated, but have their violence reduced to an acceptable level'.[39] If the state could through various means contain, undermine, damage, degrade, and erode the capacity of the PIRA, then that would perhaps generate the conditions for political progress. For terrorism to possess the Clausewitzean leverage necessary for victory, it requires its violence to be less acceptable to its enemy than it would be for the latter to accede to terrorists' demands.[40] In opposing the PIRA, the UK had long sought to produce what one 1975 secret paper described as a 'political, penal, and economic climate' favourable to the securing of the state's objectives; this combined security, political, and economic approach had at its heart the aim of changing a situation in which terrorists 'believe their ultimate political objectives can be achieved primarily by violence alone', to one in which it is shown 'that the IRA can no longer win'.[41]

The British military sometimes spoke explicitly in favour of politically integrated approaches. Leading soldier Frank Kitson's famous book *Low*

Intensity Operations (1973) emerged from the author's year at Oxford, and was aimed at making the British and other armies 'ready to deal with subversion, insurrection, and peace-keeping operations during the second half of the 1970s'. Kitson, who himself spent time in Northern Ireland, was clear that, 'as the enemy is likely to be employing a combination of political, economic, psychological, and military measures, so the government will have to do likewise to defeat him'. Kitson's own success in gaining and holding the allegiance of the population in the North was in practice highly limited. But in his recognition of the need for such support, and his advocacy of an integrated political-military approach to counter-terrorism, he was accurate enough, and was emblematic of what became a longstanding UK approach.[42]

While there exist differing views regarding the extent of continuity among UK governments in relation to Northern Ireland during the Troubles,[43] there were some clear strategic continuities of approach,[44] and these included the desire to minimize the lethal effectiveness of terrorist groups (especially the PIRA). The UK's own military assessment of the Troubles suggested that, 'The general line of strategic direction from Whitehall appears to have been to resolve matters in Northern Ireland at reasonable cost and without undue distraction to the normal conduct of business.'[45] If terrorist violence could be limited to levels at which near-normality was sustained, and at which terrorism lacked the decisive power that it required for the terrorist's own victory, then partial strategic success for the state might be achieved.

During the Northern Ireland peace process there is no doubt that ending terrorist violence was the central goal for UK Prime Minister Tony Blair and his chief negotiator Jonathan Powell. In Blair's words, 'The big prize was plainly an end to violence'.[46] Jonathan Phillips was political director (2002–5) and Permanent Secretary (2005–10) at the Northern Ireland Office (NIO): 'from the perspectives of the governments it was always a central motivating factor that the conflict had to be brought to a resolution in order to save lives. That was the fundamental and it was a powerful motivator.'[47] Getting violent groups to stop their violence, and persuading people to turn away from terrorism, were central aims of the UK state in relation to Northern Ireland. In late 1989, UK Secretary of State for Northern Ireland Peter Brooke had observed that it was hard to foresee the PIRA being militarily defeated, but that the security forces could none the less 'operate a policy of containment', allowing normal life to continue, and helping to persuade terrorists that their violence was not worth pursuing.[48]

Indeed, for most of the Troubles, the UK (including the British Army) recognized that fully defeating the PIRA in a military sense was likely to prove unattainable, that military means alone would not bring an end to the Troubles, and that including Irish republicans in a Northern Irish endgame would be necessary.[49] What they also needed (and what until the 1990s they lacked) was a situation in which the PIRA accepted the consent principle for Northern Ireland, and were prepared to end their violence without an end to UK sovereignty over the North being promised or secured. This had not been true in 1972 or 1975, nor through the 1980s. It did become true during the peace process of the 1990s and early 2000s, through a collaborative and negotiated peace settlement, centrally made possible because of this major change in PIRA approach.

If this goal, and its achievement through the peace process, are indeed judged a partial strategic success for the UK state, then part of that involves the eventual outcome in Northern Ireland aligning with what the state had long sought politically. I think that a very strong case can indeed be made here.

The peace process was crucial to this. That process involved lengthy secret communication[50] followed by more public negotiations, ultimately resulting in a deal which received impressive endorsement across Ireland. The Belfast/ Good Friday Agreement of 10 April 1998 sought to address the grievances of rival communities: nationalists and republicans who thought Northern Ireland unfair and illegitimate; and unionists and loyalists who sought to avoid being expelled from the UK against their wishes. Under this deal, power-sharing would be implemented in Northern Ireland, complemented by north–south and east–west dimensions. Important commitments were made regarding reform, human rights, and equality in the North itself. The deal therefore represented a sophisticated way of trying to address the root causes which lay behind rival terrorisms in Northern Ireland. For the UK state, counter-terrorism involved a deep commitment to politics, and rightly so. In referendums on the Agreement, simultaneously held North and South on 22 May 1998, there was striking popular endorsement: 71 per cent voted for it in Northern Ireland; 94 per cent voted for the deal in the Republic.

This involved securing much that had long been advocated by the UK as being important for a settlement. The most significant aspect of its success was this: the PIRA's long-pursued goal of Irish national self-determination now became harmonized with the Northern consent principle, a principle which the UK had long stressed as essential. The December 1993 UK-Irish

Downing Street Declaration, the February 1995 British-Irish Framework Documents, and the April 1998 Belfast/Good Friday Agreement, all saw the UK and Irish governments accept that it was for the Irish people to determine their own political future; but all also saw them stipulate that this determination would require majorities both in Northern Ireland and in the twenty-six-county Irish state if constitutional change were to occur. So the 1998 GFA settlement involved endorsement (including Irish nationalist and republican endorsement) of the principle of consent; it was now accepted that a change in the constitutional status of the North would require majority approval there. The importance of the principles that are involved here is such as to deserve substantial direct quotation.

The 1993 Downing Street Declaration:

> The Prime Minister, on behalf of the British government, reaffirms that they will uphold the democratic wish of a greater number of the people of Northern Ireland on the issue of whether they prefer to support the Union or a sovereign united Ireland. . . . The British government agree that it is for the people of the island of Ireland alone, by agreement between the two parts respectively, to exercise their right of self-determination on the basis of consent, freely and concurrently given, North and South, to bring about a united Ireland, if that is their wish. . . . The Taoiseach, on behalf of the Irish Government, [considers that] it would be wrong to attempt to impose a united Ireland, in the absence of the freely given consent of a majority of the people of Northern Ireland. He accepts, on behalf of the Irish government, that the democratic right of self-determination by the people of Ireland as a whole must be achieved and exercised with and subject to the agreement and consent of a majority of the people of Northern Ireland.[51]

The 1995 Framework Documents:

> the British government recognise that it is for the people of Ireland alone, by agreement between the two parts respectively and without external impediment, to exercise their right of self-determination on the basis of consent, freely and concurrently given, North and South, to bring about a united Ireland, if that is their wish; the Irish government accept that the democratic right of self-determination by the people of Ireland as a whole must be achieved and exercised with and subject to the agreement and consent of a majority of the people of Northern Ireland.[52]

The 1998 Belfast/Good Friday Agreement:

> The participants . . . recognize that it is for the people of the island of Ireland alone, by agreement between the two parts respectively and without external

impediment, to exercise their right of self-determination on the basis of consent, freely and concurrently given, North and South, to bring about a united Ireland, if that is their wish, accepting that this right must be achieved and exercised with and subject to the agreement and consent of a majority of the people of Northern Ireland.[53]

This approach—respecting the consent principle for the Northern Irish majority—had long been rejected by the PIRA. But it was a principle now endorsed overwhelmingly by Irish nationalists North and South in 1998, and it was to be accepted by the republican movement as the basis for an end to the PIRA's campaign.

This was all in harmony with what the UK government had long argued. The consent principle for the North had been emphasized by the British from the earliest phase of the conflict.[54] It had been endorsed by the UK's 1973 Northern Ireland Constitution Act, and had been formally accepted by the Irish government in the Sunningdale Agreement of that same year[55] (which had also sought to establish Northern Ireland power-sharing with a cross-border dimension). Again, the 1985 Anglo-Irish Agreement had seen both London and Dublin formally endorse the principle that constitutional change for Northern Ireland would require majority approval there:

> The two governments (a) affirm that any change in the status of Northern Ireland would only come about with the consent of a majority of the people of Northern Ireland; (b) recognize that the present wish of a majority of the people of Northern Ireland is for no change in the status of Northern Ireland; (c) declare that, if in the future a majority of the people of Northern Ireland clearly wish for and formally consent to the establishment of a united Ireland, they will introduce and support in the respective parliaments legislation to give effect to that wish.[56]

So the UK and Irish governments had long agreed on what they continued to stress in the 1990s peace process, that Northern Ireland could only leave the UK if a majority there wanted that to occur, and that the broad shape of the peace deal should align with architecture long promoted by these two states.

It was not, of course, that there had been a smooth or unanimous approach by the UK state throughout the conflict. Prime Minister Margaret Thatcher (in office 1979–90) had been primarily focused on issues of security against terrorism (perhaps unsurprisingly, given the terrorist killing of her colleagues, and the PIRA's attempt to murder the Prime Minister herself). She also moved in and out of having particular interest in Northern Ireland, tending to see the place as something of a distraction from more important work. But even Mrs Thatcher displayed a capacity for flexibility on policy, and her initiatives (particularly the 1985 Anglo-Irish Agreement, but also her authorizing of direct contact between the British state and the PIRA) did form an important part of the long peace process in the North.[57]

The most significant shift in the 1990s peace process was not made by the UK, but rather by the PIRA. In contrast to their own long-held view, they now accepted the consent principle for the North as part of a settlement through which they would end their violent campaign. London, Dublin, unionists, and the SDLP had all long agreed this consent principle. What was new in the late 1990s peace deal success was that paramilitary organizations and their associated politicians (most decisively, the PIRA and Sinn Fein) now agreed too. In partnership, and through multiple forms of political as well as security activity, the UK had managed to oversee an arrangement in which the kind of outcome that they had long preferred now came to be endorsed by people across Ireland. In terms of Northern nationalism, the 1998 deal was far more aligned to what non-violent actors such as the SDLP had favoured than it was to the long-proclaimed goals of violent groups such as the Provisional IRA.[58] It is also worth remembering just how hostile, during the conflict itself, the SDLP and the PIRA had been to each other.[59]

Tactical counter-terrorism (frequently flawed though it regrettably was) did play a part in producing the stalemate within which such political compromise could emerge. And the hard-headed decisions made by terrorist organizations who were being, not defeated, but substantially undermined and contained, were the crucial pivot on which this process turned.

For the PIRA, what was agreed in accepting the 1998 GFA represented a significant and brave political departure from what they had long espoused and stressed as essential to a lasting end of their violence. The PIRA had for most of their campaign interpreted Irish national self-determination as involving Ireland deciding its future as one unit, and there being no majority consent principle for the North.[60] They had explicitly linked Irish self-determination to the securing of independent and united sovereignty for Ireland,[61] and to the promise of withdrawal of British sovereignty from the island. Indeed, republicans had repeatedly stressed (privately as well as publicly) the importance of a British commitment to the latter if the PIRA's campaign were to be brought to an end.[62]

For years, the promise of an end to British sovereignty over the North had been presented by republicans as crucial to achieving an ending of the PIRA's campaign, and they had seen Northern Ireland as irreformable and needing to be removed, not amended under British rule.[63] Speaking of 1973 and of PIRA leaders of the period, Belfast PIRA man Gerry Bradley observed pithily that: 'Guys like Billy McKee or Seamus Twomey or Ivor Bell wouldn't accept anything less than a British withdrawal or a declaration of intent to do so.'[64]

It is true, and well recognized, that the PIRA considered options other than a unitary Irish state, after British withdrawal.[65] But it is also true that they long required UK commitment to withdrawal if their campaign was to cease. Leading Provisionals in the 1970s had considered British withdrawal—not just troop withdrawal, but British disengagement from Ireland, or at least the UK's commitment to such disengagement—to be an essential and achievable part of a lasting end to the conflict.[66] This is reflected in many first-hand sources. Gerry Bradley, speaking of the 1971 PIRA: 'We genuinely believed we could beat the Brits this time. . . . I genuinely believed that one day the IRA would be chasing the British Army down to the docks, firing at them, and the last British officer would be backing up the gangway onto the boat with his pistol in his hand.'[67] In the words of leading Provisional Ruairí Ó Brádaigh, 'we shall win, because we regard British disengagement from Ireland now as inevitable'.[68]

There can therefore be no doubt about the PIRA's strategic goal of removing British sovereignty over the North, establishing an independent Ireland, and ending Irish partition. When the organization formally ended their armed campaign in 2005, this central objective had not been secured. In previous phases of the Troubles they had insisted that their violence would continue unless they were assured that this central goal would be delivered.[69] Now, however, they had concluded their armed struggle and endorsed an outcome which clashed with much that they had previously presented as essential. Indeed, during the negotiations which ultimately produced the 1998 GFA, republicans seem not to have raised the issue of British withdrawal,[70] something which would have been unimaginable in the 1970s or 1980s.

The PIRA had thought an end to British sovereignty and an end to partition achievable. They had therefore pursued violently and politically something more than they ultimately settled for in 1998, and more than they had secured by the time of their final declaration of an end to their campaign in 2005. Indeed, for some PIRA members, the eventual outcome of the Troubles represented a stark departure for the republican movement from its goals and ambitions during the campaign. In the words of one former volunteer, 'To what extent did the IRA achieve their strategic goals? I would have to say they failed dismally'; the PIRA had pursued:

> the strategy of the long war, until such times as the British conceded a declaration of intent to withdraw from Ireland, and this was not some woolly concept dreamed up by a couple of individuals: it was a strategic goal widely discussed within the Republican Army and was accepted throughout the Army as the strategic goal, which was not negotiable under any circumstances. Taken from that viewpoint, the IRA completely failed to achieve its strategic goal, which was to force a British declaration of intent to withdraw.[71]

For republicans, it was important that the UK during the peace process acknowledged the right of the Irish people to decide their future unhindered. In some ways, however, this recognition changed little in practice, allied as it now firmly was to the Northern consent principle. For it had long been the case that, if the Northern majority had wanted to leave the UK, then the UK would not have stood in the way of that. Indeed, PIRA prisoners in the H-Blocks at the Maze Prison themselves lucidly made this point when the 1993 Downing Street Declaration was issued:

> we are asked to recognize as some great concession to the principle of national self-determination the British government position that should a majority in favour of Irish unity emerge in the North then Britain would not stand in its way. As if Britain could but do anything else![72]

And they were right. In November 1971, UK Home Secretary Reginald Maudling had clearly and publicly stated that, 'if, by agreement, the North and the South should at some time decide to come together in a united Ireland, if, by agreement, this should be their wish, then not only would we not obstruct that solution but, I am sure, the whole British people would warmly welcome it'.[73]

What happened in the peace process was that the republican movement recognized the stalemate that obtained in the conflict, and pragmatically decided 'to come down a few rungs'[74] (along with others) in order to achieve a compromise which offered more than would have been achieved by sustaining further political violence. Jonathan Powell, the UK's chief negotiator in the GFA talks, tellingly reflects of the republican leadership that, 'they had made the decision in the late 1980s that they wanted peace, and they knew approximately the terms they would have to settle on in the end and that would include the Union going on'.[75]

The republican movement would not have been able to make this shift on the consent principle, and to do so without fatally fracturing the movement, had it not been for the skill and commitment of Sinn Fein leaders such as Martin McGuinness and Gerry Adams. Once they decided that greater momentum would be maintained if PIRA violence stopped, then the route towards more peaceful politics was facilitated (and pro-state loyalists and the UK state could respond with a diminution of their own violence and militarization too).

In assessing the success or otherwise of this outcome for the UK, it is important to consider counterfactuals. Had there not been the tactical containment of terrorist groups that did emerge by the 1990s, would such a deal have seemed appealing to terrorist organizations? Again, could the 1998 outcome have been achieved much earlier? Niall Ó Dochartaigh has identified three key moments in 1972, 1975, and 1991 when the UK government and the Provisional IRA used back-channel contacts in the pursuit of peace.[76] The last of these efforts succeeded. Did the 1970s engagements embody missed chances for an earlier ending of the conflict? If one considers it feasible for the UK to have indicated an intention of withdrawing from Northern Ireland in the 1970s, and that peace could have been achieved by their following through on this Irish republican aspiration, then these were indeed moments of opportunity. But it is hard to take seriously the claim that British withdrawal would have produced anything other than far higher levels of violence in the 1970s, given the likely response of loyalist terrorist organizations to such a development at that time.[77] Tellingly, crucial political figures in Dublin during the Northern Troubles thought such a move far from feasible or attractive.[78] Rather than peace, an intensified civil war would almost certainly have developed. The missed opportunity argument would therefore have to run in a different direction.

Vital here is the question of whether the PIRA leadership in 1972 or 1975 might have accepted the kind of arrangement which the 1990s and 2000s PIRA eventually did endorse. Did 1970s PIRA leaders, therefore, think it acceptable and deliverable to achieve a permanent end to PIRA violence on the basis of an interpretation of Irish self-determination which incorporated the majority consent principle for Northern Ireland, and therefore the potential continuation of UK sovereignty over the North? As noted, this was what the 1990s process involved, as embodied in the 1993 Downing Street Declaration, the 1995 Framework Documents, and the 1998 GFA. During the 1990s this approach gained republican endorsement. If such a compromise on self-determination would not have been acceptable to the 1970s PIRA, then it is hard to see how peace could have been secured during those years. For the PIRA would have continued their campaign while the UK remained committed to the consent principle; and far greater violence would almost certainly have arisen had the UK decided instead to disengage from the North.[79]

The eventual PIRA peace-process position was very different from what the organization had seemed (even in private contacts) to consider acceptable in the 1970s.[80] Indeed, if the 1970s PIRA leadership had thought it acceptable and possible to end their violence on the basis of the consent principle for the North (and therefore a potentially enduring British sovereignty in Northern Ireland), then, given their contemporary back-channel contacts with the UK, it would have been easy for them to signal privately the possibility of greater compromise than was visible at the time. Had the 1970s PIRA leadership thought it appropriate and achievable to end their violence on the basis of the consent principle, then it made no sense for the organization to continue their campaign for so many painful years. They could have traded (as 1990s Irish republican leaders later did) the ending of their violence for significant reforms, within the framework of respect for the Northern consent principle. Republican attitudes to what was achievable through violence had changed by the 1990s, and it was this that was so central to making the Good Friday Agreement possible.

It was not merely through operational counter-terrorism that this progress was pursued. Echoing what has occurred in other settings,[81] the UK authorities in Northern Ireland attempted to deal with the conflict's violence by engaging with a wide range of governmental responses: economic, social, and political, as well as more directly counter-terrorist work.[82] This security and military aspect of things was substantial, however, and the vital role played by intermediaries, interlocutors, and negotiators in the peace process[83] should not obscure that. The complex dynamics are valuably reflected in the words of Quentin Thomas, political director in the Northern Ireland Office during the crucial years 1991–8:

> The key question was whether the republican terrorists, and in particular the PIRA, continued to believe that their campaign could succeed. It was the singular success of the security forces and the intelligence community, coupled with the robust position taken by the main political parties in the United Kingdom, to deny the republicans belief in their prospective victory.[84]

The UK's analysis of Operation Banner (the name given to UK armed forces' operations in Northern Ireland between August 1969 and July 2007) noted that at the height of the campaign in 1972 there were 28,000 British soldiers in Northern Ireland; that, in total, more than 250,000 regular army soldiers served in Northern Ireland during the Troubles; and that more than 600 soldiers had died as a result of terrorist violence. It also noted, however, that much of Northern Ireland during much of the Troubles was 'relatively peaceful', and that violence varied greatly according to location. Of the Provisional IRA, the UK's own assessment judged that, 'PIRA developed into what will probably be seen as one of the most effective terrorist organizations in history. Professional, dedicated, highly skilled, and resilient, it conducted a sustained and lethal campaign in Northern Ireland, mainland United Kingdom (UK) and on the continent of Europe.'[85]

Tactically, much was done to limit and contain the PIRA, and we will consider the complexities of that shortly. But if the Northern Ireland peace process represented a means of ending terrorism and generating partial strategic victory, then the dynamics of that wider process must form part of our reflection on UK counter-terrorism in Northern Ireland.

Relationships were vital in the contingent process of making peace in Northern Ireland. The commitment of high-ranking politicians (John Major, Tony Blair, Albert Reynolds, Bertie Ahern, Bill Clinton, George Mitchell) together with some of their colleagues (Blair on Jonathan Powell: 'Without him, there would have been no peace')[86] was striking.

In the UK's own retrospective assessment:

> Security forces do not 'win' insurgency campaigns militarily; at best they can contain or suppress the level of violence and achieve a successful end-state. They can thus reduce a situation to an 'acceptable level of violence'—a level at which normal social, political, and economic activities can take place without intimidation. 'Acceptable level of violence' as a term should be used carefully since violence should have no place in a developed society. What is required is a level which the population can live with, and with which local police forces can cope. Security forces should bring the level of violence down to the point at which dissidents believe they will not win through a primarily violent strategy and at which a political process can proceed without significant intimidation.[87]

Again, of the British Army, the UK's own assessment was:

> It should be recognised that the Army did not 'win' in any recognizable way; rather it achieved its desired end-state, which allowed a political process to be established without unacceptable levels of intimidation. Security force

operations suppressed the level of violence to a level which the population could live with, and with which the RUC and later the PSNI could cope. The violence was reduced to an extent which made it clear to the PIRA that they would not win through violence.[88]

The idea that the Northern Ireland peace process was built on the foundation of a stalemate is one that has received divergent scholarly analyses.[89] But the evidence is clear that, in military terms, there was indeed a stalemate (a deadlock which allows neither side to win), and that this was certainly recognized by sharp-eyed republicans.[90] This stalemate made it more appealing to terrorist organizations to explore alternative means of pursuing their goals. The political relationships, opportunities, compromises, and possibilities that emerged down this alternative route made more sense to republicans than a continuation of rather futile (and partly contained) violence.

Partial strategic victory for the UK state did not therefore involve defeating the PIRA. It did involve the state showing resilience and an enduring commitment. And it did involve containing the Provisionals in such a way as to make peace-process politics seem more attractive than terrorism. Relatedly, it involved preventing the PIRA's own victory through sufficiently thwarting their operations and substantially containing their campaign.[91] Republican responses here owed much to the fact that the PIRA was a very pragmatic organization, and one which would pursue what it believed it could achieve.[92] There is a very strong and admirable pragmatism within Irish nationalism,[93] and republicans' seizing of alternative opportunities through the peace process reflects this long tradition. If violence would not secure victory, then other forms of politics might prove more fruitful. In Sinn Fein's Conor Murphy's thoughtful phrasing, 'If it was becoming fairly clear that armed struggle wasn't going to drive people onto boats and out of Belfast harbour, then you had to have another dynamic which opened up the possibility of great political change'.[94]

A preparedness to alter tactics, and to act in the most likely way to secure political momentum, reflected a lengthy tradition within nationalist Ireland of complex relationships between violent and non-violent politics. The boundary between violent and non-violent traditions here has long been porous, with attitudes and tactics being marked by considerable fluidity,[95] and with individuals and organizations shifting approach as circumstances seemed to require. Indeed, even those whose commitment to extremist politics is read by many in terms of simplistic thinking can turn out, on close inspection, to be complex figures.[96]

This gives context to the suggestion by Joseph Pilling (permanent secretary at the NIO during 1997–2005), that 'it is rather questionable that we would ever have got as far as we did in political talks if Sinn Fein had believed that the Provisional IRA were likely to achieve their goals by shooting and bombing, and get the Brits to leave and achieve a united Ireland that way'. Crucial here had been the attempt 'to convince republicans not that they would lose the military campaign that they were engaged in, but that it was not a campaign that they had any particular prospect of winning'.[97] Tactical containment and the thwarting of terrorist organizations were vital to this. In particular, the PIRA pragmatically settled for something very far from what they had long pursued through violence, recognizing the reality that victory was not emerging.[98]

Beyond this partial achievement of central strategic goals, a case could also be made that the UK state managed to secure the secondary goal of reassuring the public that the state would endure and that order maintenance would be upheld. The UK's own assessment of its record in Northern Ireland claims that, 'The events of 1969 could easily have turned into open civil war, but did not.'[99] The state clearly considered this a major accomplishment, and (despite the awfulness of the violence, especially at its high levels in the early years of the Troubles)[100] this perception was not without foundation. The UK state did show a preparedness for long commitment in Northern Ireland, and it developed sufficient long-term resilience to demonstrate its capacity to survive terrorism.

More ambiguous would be claims about partial strategic success in terms of determining the agenda. It is true that the UK state prevented anti-state terrorists from securing victory through violence. It is certainly not true that the existing borders of the UK state are securely established for a long future. The 2016 Brexit referendum contributed to a greater momentum behind nationalist separatism in both Scotland and Northern Ireland, although (in terms of counter-terrorism) the state could fairly claim that its work had been done if it had limited the degree to which separatists in Northern Ireland now deployed violence. Most people in the North have decided to eschew support for terrorist groups, one major legacy from the Troubles being that most people do not judge terrorism to have shown itself the best or only way of pursuing important political objectives.

In other ways, however, militant actors have still managed to sustain and transmit a sense that their own arguments had been and remained legitimate.[101] There proved a durability to the political narratives on which violent movements had relied, even after their own violence had been contained and largely come to an end. Around a decade after the 1998 GFA, there was evidence that former prisoners associated with paramilitary organizations still very much held to their previous political analysis and that, among many of them, there had been little renunciation of prior beliefs, ideological orientation, or political commitment. Rather, pro-state loyalists explained their 1990s cessation of violence on the ground that they had won the war, while anti-state republicans argued that they would eventually secure their long-held objectives, but that largely non-violent means would now best achieve these. In other words, there had been a change in tactical approaches towards the need for violence, rather than any sea change in central political narrative, opinion, or ideological objectives. People had decided now to refrain from violence, while retaining the commitments, goals, and views that had initially been seen as justifying it, and indeed while adhering to the view that that prior violence had been legitimate, just, necessary, and productive. Moreover, each community's former prisoners substantially retained their earlier and largely negative view of the other community, of the enemy.[102] Peace-process engagement by the PIRA was therefore practical and pragmatic. Moreover, ex-paramilitaries (especially on the republican side of Northern politics) have gained much respect, status, and influence within their communities through their having been involved in violent struggle;[103] and the UK state failed to produce a shared narrative which both communities could simultaneously espouse.

5

The Dynamics behind Peace

I

The UK state in Northern Ireland secured a political deal which largely achieved peace. It did so in line with frameworks which the state had long proposed, and which had previously been rejected by major terrorist organizations and their politicians. There were many reasons for this partial strategic success, and tactical counter-terrorism formed only one part of the process. The UK showed a willingness to compromise, whether in its utilization of the language of self-determination, the serious reforms of Northern Ireland that were involved in the peace process (in regard, for example, to the police), or the concessions given to terrorist groups in relation to prisoner release. The Irish government likewise made their own concessions to help secure unionist and loyalist support for the GFA. The Irish state's claim to the territory of the North was significantly amended, for example, to acknowledge that the achievement of Irish unity would require majority consent in Northern Ireland.[1] There was agency on all sides, as the UK and Irish governments, international actors such as the US, and paramilitary groups all made decisions and concessions amid pragmatic dialogue towards a compromise settlement.

But part of the rationale for terrorist groups (most significantly the PIRA) to opt for this lengthy process of remarkable compromise was this: their violence could no longer credibly be considered to be leading to the victory that they had long believed possible. Indeed, this echoes a wider reality of terrorist violence. While the logic of such methods relies on the idea that they will become increasingly unbearable for one's opponents, and that they will therefore lead to success, the more normal pattern is that the state becomes better able to contain (and prove resilient in the face of) terrorism. Even in terms of publicity and the pressure generated by shocking violence, populations and governments become *more* rather than less able to endure. This was partly what undermined the PIRA's hope that British opinion would, in response to republican violence, increasingly pressurize London governments towards the granting of Irish republican demands.[2]

The tactical aspects of UK counter-terrorism therefore played a crucial part in limiting terrorist capacity, containing political violence, and undermining the possibility of victory for terrorist groups. This is not to suggest that the UK defeated the PIRA (an exaggerated simplification), nor to propose that the Good Friday Agreement emerged primarily because of the work of the Royal Ulster Constabulary (RUC), MI5, or the British Army. It is rather to point out that the work of those and other wings of the UK state did limit terrorist capacity in ways without which the dramatic change of direction by terrorist groups (most significantly by far, the PIRA) is inexplicable. As we will see, UK tactical counter-terrorism was far from flawless and could involve seriously counter-productive elements. But tactical achievements did help to create one of the conditions necessary for the UK's partial strategic success in Northern Ireland.

There were certainly many tactical-operational successes. By the later years of the conflict the authorities were preventing many terrorist operations from coming to fruition. Much of this was based on the use of intelligence. In the words of Gerry Bradley (a Belfast PIRA man during 1971–94): 'After '76, the number of ops dropped dramatically. You couldn't have an op every day. The Brits knew all the tricks by then and who to look out for.... They knew who they were looking for.... All the players were marked by the mid-eighties. The Brits noted where they were seen, where they were coming from, and where they were going.'[3] One former RUC Special Branch (SB) officer suggested that about 60 per cent of Troubles intelligence was gained from human sources, 20 per cent from technical means such as eavesdropping or telephone intercepts, 15 per cent from surveillance, and 5 per cent from either routine patrols or open sources such as local newspapers.[4]

This work reflected a learning experience, as the UK became more adept at the gathering and utilization of intelligence over the painful course of the Troubles. The British military's own judgement is that intelligence had been limited as the conflict commenced: 'There was very little actionable intelligence before the introduction of internment. The RUC SB was almost completely ineffective and the traditional source of HUMINT [Human Intelligence]—the B Specials—had been disbanded.' But things improved dramatically in subsequent years: 'The quality of intelligence became very good indeed—by the end of the 1980s PIRA was unable to mount a bombing operation in Belfast for about two years.'[5] Terrorists' capacity to kill British soldiers declined markedly, if one compares the 1970s with the 1980s, for example.[6] And large numbers of terrorist attacks were certainly being prevented. Former PIRA member Eamon Collins suggested that, from his own experience, no more than 50 per cent of PIRA operations came to their planned fruition, though state counter-terrorism was only part of his explanation for this: 'There were many reasons for the low success rate, including failure of nerve on the part of volunteers, interception by Crown forces, sheer incompetence, betrayal by informers, and simple bad luck.'[7]

Special Branch was the RUC's intelligence unit, and was central to the state's intelligence work in Northern Ireland. By the mid-1980s the RUC was paying more than £500,000 per annum to informers,[8] and SB developed considerable expertise at blocking PIRA operations, not least through this use of informers and the information they provided.[9] RUC SB E3 was the department that ran agents; RUC SB E4 did surveillance; E4A (created in 1976) was an intelligence-gathering unit; and E4B (Headquarters Mobile Support Unit, or HMSU, established in 1980) was the tactical support or kinetic police unit, doing complementary and more muscular special operations. While the UK Security Service MI5 took over the lead role in Northern Ireland intelligence and security work in 2007, for much of the Troubles it was RUC SB that represented the centre of gravity of UK counter-terrorist intelligence. Some former Special Branch detectives suggest that every full PIRA member who was a double-agent saved nearly forty lives a year; former RUC SB officer William Matchett suggests that agents in republican paramilitary groups alone saved more than 16,500 lives during the Troubles.[10]

The British Army too operated intelligence-gathering units. The Military Reaction Force (MRF) operated undercover in the early 1970s in Northern Ireland, and comprised approximately thirty-five soldiers drawn from the British military. Its aims were to gather intelligence but also to operate in a more aggressive fashion as a counter-terrorist unit combatting the IRA, and frustrating the latter's operations whenever possible. The work could be brutal. One former MRF soldier recalls the unit being encouraged to 'deal with' and 'eliminate' the terrorist enemy, to identify and destroy particular terrorist opponents.[11]

Subsequently, the Force Research Unit (FRU) was a British Army Intelligence Corps working in Northern Ireland, recruiting, running, and utilizing the Army's human intelligence sources in the counter-terrorism war there. Based in the Army's Lisburn Headquarters, the FRU operated from 1980 until the early 1990s, when it became renamed the Joint Services Group. The FRU's work could have tactical-operational but also politically related benefits. Derry's Willie Carlin, for example, worked for MI5 and for the FRU as an agent within the republican movement: the Sinn Fein activist was also a British operative and, as such, provided information and insights about republican political developments and personnel during a fourteen-year role.[12]

There was therefore a clear process of improvement in UK intelligence work during the Troubles, from a self-damagingly uncoordinated early phase[13] towards a more sophisticated and successful means of containing terrorist opponents. Indeed, the Troubles became an intelligence-led war. For the state, as suggested, much of this involved the work of agents and informers. Considerable evidence points to the PIRA having been significantly infiltrated by the UK and Irish states, not to the extent of defeating the Provos, but certainly to a degree which hampered that organization's ability to achieve victory.[14] One former RUC officer claims of the late-Troubles PIRA that, 'They were being contained. . . . It became stalemate.'[15] Informers were important here ('Agents were the decisive factor in the intelligence war'),[16] but so too was technical surveillance. The latter, in the words of one ex-HMSU officer, 'was massive. It was absolutely crucial to it.'[17]

What did such methods mean in practice? On 8 May 1987, the SAS ambushed a PIRA team as they attacked an RUC station in Loughgall, County Armagh. All eight PIRA members were killed, as was a civilian who happened to be nearby. The authorities had learned of the PIRA attack through a surveillance device located in premises used by a republican.[18] The incident understandably became controversial. The SAS had given no warning before opening fire, although it is also true that the PIRA clearly were in the process of murderously attacking police officers, and in the months following the Loughgall attack there was a diminution in terrorist activity in the border areas.[19]

Some of the most effective counter-terrorism work in this field was done by Northern Irish figures. Ian Phoenix (1943–94) was a Detective Superintendent in RUC Special Branch, focusing on top-secret counter-terrorist surveillance and undercover operations, especially against the

PIRA. A Northern Ireland-born former soldier, his work and that of his Special Branch colleagues involved the gaining of extensive intelligence through informers and through human and technological surveillance.[20] But other work was done by non-Northern Ireland counter-terrorists. Some of the latter-day intelligence successes against the IRA involved the foiling of major plots through work led by MI5: during the twelve-month period up to July 1994, eighteen of thirty-four PIRA operations planned for Britain were prevented; other 1990s tactical-intelligence successes included the disruption of a potentially baneful PIRA plot to attack the power supply for the whole of greater London, and the prevention of a major bombing attack on central London.[21]

It is important to note that there was a long-historical quality to this aspect of the modern Troubles. English and British authorities had long utilized informers in their attempts to undermine Irish republican movements, and the informer is indeed an anciently rooted phenomenon. There could be a complexity of motives for those who decided to betray their comrades. These variously included: greed and financial gain; the desire to avoid prison; excitement and thrill; a keenness to avoid exposure to their colleagues by the authorities for some transgression; disillusionment with the cause or with the terrorist organization; a sense of guilt (especially around particularly egregious operations), and a desire to redeem oneself through life-saving, altruistic work; a falling out with people within one's organization, and perhaps a desire for some revenge upon them; a keenness to be respected, approved of, valued, taken seriously, and seen to be important.[22]

If some became agents because of disillusionment with the terrorist cause, it is also true that agents' work could itself generate further disenchantment. The presence of traitors in an organization could undermine morale among informers' colleagues, creating doubts about others' loyalty, and cumulatively dampening people's belief in possible victory. The individuals involved in this dangerous work could therefore have very significant effects.

Sean O'Callaghan (1954–2017) was a Kerry-born member of the PIRA, who came to view 'the birth of the Provisional IRA as the greatest tragedy in modern Irish history', and who 'realized that joining the Provisional IRA had been the biggest mistake of my life'. O'Callaghan had been a teenage recruit, and the republican movement had become his life. But then he turned decisively against it: 'As time went on it became more clear to me that the Provisional IRA was the greatest enemy of democracy and decency in Ireland.' He came to recognize and revile the sectarianism of the PIRA:

'Helping to inform democratic governments about IRA terrorism and possibly helping to save lives seemed to me to be a decent thing to do'; he wanted 'to help damage the killing capacity of the IRA'.[23] O'Callaghan became a significant informer for the Irish police.

Eamon Collins (1954–99) was another Provisional IRA member who came to develop a profound disenchantment with what he judged the PIRA in practice to be doing: the counter-productively divisive effects of PIRA violence; the flawed motivations and character, and the callous brutality and sectarianism, of some of his republican comrades; and the naivety of PIRA political analysis. He also felt considerable guilt for his own actions, and for a time became an informant for the RUC: 'I am now deeply hostile to the IRA, and look forward to its demise.'[24]

Freddie Scappaticci grew up in the Market area in Belfast. The son of an Italian immigrant, he was interned in 1971, released in 1974, and became a British Army agent (codenamed Stakeknife) within the PIRA in 1978. By 1980, he was working in the PIRA's internal security, counter-intelligence department (referred to as the 'Nutting Squad' because of its practice of shooting people in the head—'nut'—if they were judged to have been traitors). The Nutting Squad's duties included investigating suspected informers and inquiring into operations that appeared to have been compromised. Scappaticci's decision to become an agent seems partly to have derived from his having held hostile attitudes towards some other PIRA figures, and to his having been beaten by republican colleagues after an argument about policy. It appears that Scappaticci's information did help to stop some PIRA missions.[25] But it also seems clear that Scappaticci himself continued to act in brutal, murderous fashion as a PIRA man, while also working for the British Army, and indeed that the UK state allowed other people to be killed in order to protect his identity as an agent.[26]

The costs of turning against your former comrades could be high. Eamon Collins was brutally murdered in 1999, and the case of Denis Donaldson provides another grisly example. A Sinn Fein and PIRA activist, Donaldson had become an agent for MI5 and the Police Service of Northern Ireland, and was killed by republicans in County Donegal in 2006 after it became known that he had been working as an agent.[27]

One of the most significant means of tactical-operational containment involved the prosecution and imprisonment of large numbers of terrorists. During the Troubles, thousands of people were jailed for terrorist-related offences, and this represented a major contribution to the limiting of terrorist organizations' capacity.[28] The impediment of terrorist logistical work was also important. On 28 March 1973, the PIRA's Joe Cahill was arrested on board the *Claudia* by the Irish navy, who had been given information by the UK authorities; the ship was carrying large amounts of weaponry and ammunition which had been provided by Libya. On 31 October 1987, the *Eksund* was intercepted off the coast of Brittany, having departed from Tripoli with 150 tonnes of Libyan-provided explosives and weapons.

The main importance of this successful tactical work was that it could in various ways contribute to strategic or partial strategic success. It seems clear that the UK used its intelligence networks to try to encourage movement towards more political and less violent modes of struggle in the republican movement.[29] It is also clear that, while terrible loss of life did continue for many years in Northern Ireland, the state did prove able to contain and reduce terrorists' lethal capacity. During the period 1970–9 the PIRA killed on average 105 people every year; during the period of the state's more effective tactical counter-terrorism (1980–93), the PIRA killed on average forty-nine people per annum.[30] Counter-terrorist tactics were not the only reason behind these figures. But, given the state's demonstrable achievement of increasingly being able to thwart the PIRA's violent attacks, they undoubtedly played a significant part in this life-saving process.

Not all aspects of the intelligence war went well for the UK, of course. Intelligence was crucial to all combatant organizations in the Troubles, and tactical work did not always redound to the advantage of the UK state. The PIRA, for example, often successfully gathered information which could identify the who, where, and when of possible terrorist targeting. As one candid former PIRA activist put it, in regard to gathering the 'little bits of information' that facilitated the targeting of victims: 'People talk and people die.'[31]

It is also true that some tactical-operational work by the state proved counter-productive. There is little doubt that the introduction of internment without trial in 1971 undermined rather than strengthened the British state's authority in Northern Ireland. Understandably concerned about the grim security situation in Ulster at that early stage of the conflict, the government opted to arrest some of those suspected of involvement in violence. But the policy was initially implemented one-sidedly against nationalists and the operation failed to deliver the desired results, not least because many innocent people were arrested, and because of the mistreatment of some of those who had been lifted. In practice, therefore, a measure intended to undermine terrorism had the effect of strengthening support for terrorist violence against the state; poor timing and preparation helped to make the measure ineffective.[32] As one British soldier later put it, internment was: 'a complete disaster...in any internal security operation—and that's what Northern Ireland was—hearts and minds are the most important part of it. And internment destroyed it.'[33] Hostility to internment was also the cause of the march in Derry on 30 January 1972, at which fourteen people were fatally shot by the British Army on what became known as Bloody Sunday.[34] The terrible human suffering on that day is its most important, profound reality. In terms of tactical-operational misjudgement and counter-productiveness, it was also a grave example of counter-terrorist failure. Collusion, literally the sharing of a secret understanding, clearly has many potential applications in relation to counter-terrorism. If the police are going to gain life-saving information from informants in terrorist organizations, then clearly some form of secret understanding is essential (and potentially benign for those whose lives or limbs are saved if attacks are blocked through the acquisition of information).[35] In the words of the Police Ombudsman for Northern Ireland in 2022: 'I...accept that often only those deeply embedded within terrorist organizations could provide the high-grade, actionable intelligence which police required to disrupt paramilitary activities, secure convictions, and prevent loss of life.'[36]

The ethical challenges here can be lastingly difficult, however.[37] A long-term informer in a terrorist organization will only be able to deliver life-saving information if they remain actively part of that organization. This means that the state must make an essentially utilitarian decision about whether the benefit derived from their information sufficiently outweighs the problem of holding a secret understanding with someone involved in illegal activity. That problem is increased by the fact that, obviously, not everything done by someone working as an informer is directed or controlled by the state operatives who are episodically receiving information from them. Moreover, collusion can involve much darker activities, such as dual membership between state forces and terrorist organizations, the active attempt to collaborate to do harm to particular citizens, or the providing by state actors of information or weaponry likely to be used by terrorists in targeting people. No justification can be sustained regarding such malign activities, and nothing can erase the terrible pain caused.

Within this complex pattern, considerable attention has rightly been paid to evidence about collusion between some people within the UK counter-terrorist community and loyalist paramilitaries.[38] The Glenanne Gang, for example, was a sectarian group comprising some UVF, RUC, and Ulster Defence Regiment (UDR) members, and which murdered victims in 1970s Northern Ireland and across the Irish border. Again, the horrific suffering inflicted by these killers is its most important aspect. But it is also true that instances like this, where some members of state forces colluded with organizations such as the UVF, clearly contributed to sustaining cycles of murderous violence rather than to diminishing them. This violence also degraded the state through behaviour which understandably demolished many Irish nationalists' confidence in the fairness of the authorities. In particular, the overlap in membership between the UDR and loyalist terrorist groups (a minority of UDR members, of course, but still a shocking reality) represented what should have been an entirely unacceptable situation.[39]

Collusion between some members of Northern Irish state forces and loyalist terrorists was as reprehensible as it was probable (given the divergent allegiances held by different community members during such a bloody and sustained civil war). The important thing analytically is to avoid an approach which merely seeks out evidence confirming one's own prior view (that the extent of collusion was negligible, for example, or that it was so pervasive as to be definitive of loyalist violence as a whole). Such views may in time prove accurate, perhaps. But proof will require a systematic appreciation of what the evidence compels us to believe, what it prevents us from believing, and what it allows us reasonably to infer. More comprehensive research needs to be done on this subject than has been produced to date. But it seems clear that there was on occasions appalling collusion with terrorists on the part of some (a minority of) RUC and UDR personnel. It also seems clear that London's primary focus on anti-state terrorism meant that the UK did less to address this problem than should have been the case (partly because it was judged that having some loyalist-sympathizing people in the communally one-sided UDR was preferable to their acting entirely within the realm of non-state groups).[40] This latter view—that even a sometimes collusive UDR acted as a safety valve on loyalist violence— seems to me to be at the unacceptable end of state pragmatism. To the extent that it was indeed held, it was surely one of the most deplorable aspects of UK state attitudes towards the Northern Ireland conflict. UDA and UVF terrorism represented just as much of an atrocity against citizens' lives and security as did PIRA terrorism. The UK state should have acted with even-handed commitment to address both kinds of violence throughout these and other terrorist organizations' campaigns.

Overall, however, research on loyalism does not support the view that pro-state terrorism in Northern Ireland can be explained, in its origins or in its blood-stained dynamics, primarily by reference to collusion by some state actors.[41] And it is important to acknowledge that most victims of loyalist terrorism were not republicans whose names and details had been leaked to loyalists; frequently, indeed, they were Catholics randomly selected.[42] The vast majority of loyalist targets had not been set up by the state; it is also true that many loyalists were imprisoned by the authorities, and that many loyalist attacks were intercepted by the state.[43] Though asymmetrical, UK policy did repeatedly and extensively involve the pursuit and imprisonment of large numbers of loyalist terrorists. Moreover, the secret understandings developed by the state with informers (in both republican and loyalist organizations) did save many lives, as well as helping to contain and limit the efficacy of major terrorist organizations.

None of this is to ignore the reprehensible quality of the worst aspects of collusion. Some have made very strong claims here. Martin Ingram is the pseudonym used by former British soldier Ian Hurst, who worked in Army intelligence in Northern Ireland and who has made some striking allegations about the Troubles. In his view, 'the British state organized and participated in state-sponsored murder' in Northern Ireland: 'The state was not just an arbitrator, a peacekeeper, it became a participant on the loyalist side.'[44]

Systematic analysis has been offered by others, but again there has emerged shocking material. Distinguished British barrister Sir Desmond de Silva led a Review into the UDA murder of lawyer Patrick Finucane in Belfast on 12 February 1989. In relation to counter-terrorist intelligence, de Silva tried to clarify something of the Troubles context:

> The submissions made to my Review by all former intelligence officers stressed that an agent could only provide the most valuable, and potentially life-saving, intelligence if they were infiltrated into the heart of a terrorist group. It followed that agents who were so infiltrated would, in order to maintain their cover, be required of necessity to engage in criminal conspiracies with their terrorist associates (whilst, in theory, seeking to help the security forces to frustrate the realization of these plans).[45]

But de Silva concluded that there had not, in the late 1980s, been an adequate legal framework and policy for agent-handling in Northern

Ireland, and that MI5, the RUC Special Branch, and the British Army's Force Research Unit had each therefore operated under separate modes of working. This led, in de Silva's view, to there being contradictory guidance and considerable ambiguity about how far agents were allowed to engage in criminal activity as they gathered intelligence for the state.[46]

Perhaps most strikingly, de Silva also found as follows:

> It was apparent that successive governments knew that agents were being run by the intelligence agencies in Northern Ireland without recourse to any effective guidance or a proper legal framework. I found that repeated attempts were made by senior RUC, Security Service and (latterly) Army officers to raise this very issue with government ministers at cabinet level. Yet it was not until 1993 that some cabinet ministers belatedly came to support the creation of a legislative framework. Even then, it was not until seven years later, when the Regulation of Investigatory Powers Act 2000 (RIPA) was passed, that any description of a statutory regime was created. . . . My overall conclusion is that there was a wilful and abject failure by successive governments to provide the clear policy and legal framework necessary for agent-handling operations to take place effectively and within the law.[47]

On the appalling murder of Mr Finucane, de Silva concluded 'that two agents who were at the time in the pay of agencies of the state were involved in Patrick Finucane's murder, together with another who was to become an agent of the state after his involvement in that murder became known to the agency that later employed him.' He also observed that, 'in my view the FRU did not have foreknowledge of the conspiracy within the UDA to murder Patrick Finucane', but suggested that 'proper exploitation' by RUC Special Branch of intelligence that they had prior to the attack could indeed have prevented the murder of Finucane.[48] In particular, it is surely outrageous that the quartermaster (the provider of weapons and material) for the Finucane murder (William Stobie) seems to have told his Special Branch handlers that a murder was planned, and where the weapons were to be stored, and that those weapons were not intercepted or disabled.[49]

More broadly, during the late 1980s, in Desmond de Silva's words, 'It is clear that there were extensive "leaks" of security force information to the UDA and other loyalist paramilitary groups. Many stalwart individuals served in the security forces during this time and my conclusion should not be taken to impugn the reputation of the majority of RUC and Ulster Defence Regiment (UDR) officers, who served with distinction during what was an extraordinarily violent period. Nevertheless, it is clear that some individuals within those organizations provided assistance to loyalist paramilitaries in instances where they shared a common desire to see republican paramilitaries killed. Such leaks were not institutional nor systemic, though they could certainly be described as widespread.'[50]

Brian Nelson was an agent for the FRU during the 1980s. During the time that he was working for the Army, he was involved in murders and attempted murders of Irish republicans. The Army clearly failed to supervise Nelson's activities appropriately and, on occasions, his handlers even seem (in de Silva's words) to have 'provided him with information that was subsequently used for targeting purposes. These actions are, in my view, indicative of handlers in some instances deliberately facilitating Nelson's targeting of PIRA members.' Moreover, the RUC Special Branch clearly seem to have failed to act on intelligence provided by Nelson about impending attacks.[51]

Complementing de Silva's Review, state-sponsored assessments of potential state wrong-doing have continued well into the twenty-first century. A Report in 2022 by the Police Ombudsman for Northern Ireland (into the handling by police of certain loyalist murders and attempted murders during the late 1980s and early 1990s) judged that in these years there was evidence of some 'collusive behaviours on the part of police', and that 'the significant amounts of documentation recovered by the RUC from loyalist intelligence "*caches*" indicated that both the UDA/UFF and UVF had access to security force information for targeting purposes'. It further asserted that 'a significant number of serving and former UDR members had links with loyalist paramilitaries in the North West during the period in question. This included senior figures within the North West UDA/UFF. The infiltration of the regiment in this manner allowed paramilitaries access to weapons, training, intelligence, and uniforms which added to their effectiveness in carrying out sectarian attacks.'[52]

Significant also, however, was the Ombudsman's statement: 'I am of the view that there was no specific intelligence that could have forewarned of any of the attacks, and allowed police to have taken preventative measures'; 'This investigation has not established that any police officer committed a criminal offence by protecting an informant from arrest and/or prosecution. On the contrary, my investigators identified a number of occasions where informants were arrested and reported to the Director of Public Prosecutions (DPP) for prosecution.' Indeed, the Ombudsman (Marie Anderson) stressed: 'I am of the view that the majority of RUC investigative actions, in relation to the attacks outlined in this public statement, were progressed in a thorough and diligent manner. Evidence was gathered and my investigators established that attacks were linked and considered as a series, as opposed to isolated incidents. The majority of intelligence obtained by RUC Special Branch was shared with murder investigation teams in a timely manner. I have found no evidence that RUC investigators sought to protect any individual from prosecution. Arrests were made and, where evidence existed, files submitted to the DPP. A number of individuals were prosecuted and convicted.'[53]

Lengthy quotation from Desmond de Silva and from Marie Anderson is important because the pictures that they paint are not as straightforward as some people have presented them to be. Very serious questions are raised here about the actions of the UK state in Northern Ireland; but appreciation of context and of the proportion of the wrongdoing are also evident and significant.

Part of the problem, where things did go tactically wrong for the state, perhaps emerged from the fact that the RUC were indeed refused appropriate guidance from London on what was and was not permissible in terms of agents and their activities.[54] That represents the context for some of those errors—egregious in themselves—which undermined the moral legitimacy of the counter-terrorist campaign of the police during the Troubles. And, despite the many successes achieved through high-grade intelligence and other counter-terrorist work, the evidence available too often points to a mixture of callous non-state terrorism, grisly and tragic murder, and only partial counter-terrorist efficacy.

A couple of tragic examples illustrate the point.

On 22 April 1978, off-duty police officer Millar McAllister was murdered by the PIRA at his Lisburn home. McAllister had two sons (respectively aged 11 and 7 at the time of the killing). Both boys were at home with their father when he was murdered; indeed, the younger saw his father shot dead in front of him. Millar had been identified by the PIRA as a target partly because he wrote as the Northern Ireland correspondent for the *Pigeon Racing News and Gazette*; his reports were accompanied by a photograph, they sometimes used the byline 'The Copper' (slang for policeman), and they on occasions featured his actual name. Somebody who had been brought into Castlereagh Police Station linked Millar (who sometimes worked there) with the picture that they had seen in the magazine. Revenge and hatred accompanied political motivation for the PIRA members who carried out the killing. For the terrorists, therefore, there were perhaps some inherent rewards in terms of revenge against a hated enemy, and clearly the state's protective measures had here failed to prevent the murder of one of its officers. But after the event there was some counter-terrorist success, and the identification and capture of those responsible for the PIRA killing arose partly from the work of an informer.[55]

The PIRA kidnapping and callous murder of German businessman Thomas Niedermayer (orchestrated by Brian Keenan) represents another grisly illustration of the complexities and difficulties of counter-terrorism in Northern Ireland. Niedermayer had been living in Northern Ireland since 1961 and was manager of an electronics factory on the outskirts of Belfast, one which employed people from both political communities. He was popular, fair-minded, hard-working, and not involved in the politics of the Troubles. In December 1973, he was abducted by the PIRA, with the aim of forcing the UK authorities to move two PIRA prisoners then held in England to a prison in the North of Ireland. The PIRA lied and denied responsibility for the abduction, and Niedermayer's wife and two daughters (all of whom were understandably traumatized by the atrocity, and all of whom later, tragically, killed themselves) therefore had no knowledge of what had happened to their husband and father. This process continued for years. In fact, after only a few days of captivity, Niedermayer had tried to escape, had been brutally beaten, and consequently died. The PIRA then buried his body but did not tell anyone what had happened or the where-abouts of the corpse. Eventually, an informer within the PIRA told the RUC where the body had been dumped in a shallow grave. The police had remained committed to finding the businessman's body and, in 1980, they were able to locate it and to allow there to be a funeral.[56] Such episodes—of brutal murder, compounded by years of cold-hearted silence—deserve greater prominence within the memory of the Troubles than they some-times seem to be afforded. We could also reflect that the one comforting aspect of this vile episode emerged as a result of an informer.

In some ways, these tragic stories spell out the wider truth that (unsur-prisingly) much could have gone better operationally for the state in North-ern Ireland. As the UK's own assessment of the campaign acknowledged:

> At no stage in the campaign was there an explicit operational level plan as would be recognized today....In practice much depended on individuals, their personalities, and how they got on together. Overall the picture is of

generally able and well-intentioned men doing what they believed best with a generally similar common purpose. In practice, too many things that were everybody's job were nobody's job. It could have been better.[57]

But, very significantly, the stalemate which formed part of terrorist groups' decision to pursue peace-process politics did derive partly from the state's tactical containment of paramilitary groups and their operations. Likewise, the avoidance of more full-scale civil war played its part in creating the conditions within which many elements of politics could move actors towards the 1990s deal which substantially ended the Northern Ireland conflict. As one former IRA man put it, 'If all of the IRA's planned operations had been successful, then large parts of the North would have been turned into no-go areas for the Crown forces';[58] had all planned republican and loyalist attacks come to fruition, with even further cycles of escalation, could a peace process really have developed as it did?

There are those who dispute the importance of the UK's intelligence work in bringing the PIRA towards the compromise embodied in the Northern Ireland Peace Process.[59] But it seems clear both that a military stalemate was part of what inclined the PIRA towards creative engagement with peace-process politics and also that the state's intelligence-based capacity to limit PIRA operations had contributed something to the existence of that stalemate. This is not to say that the PIRA were militarily defeated, nor to deny the many successes that they themselves achieved, the innovations that they initiated, or the tactical ingenuity that they displayed.[60] Nor is it to claim that they could no longer carry out lethal attacks at the point when they opted for their 1990s ceasefires, or that intelligence-based containment was the only reason for the republican shift towards less violent politics. But the PIRA had been an organization which saw its violence as a means of achieving a united and independent Ireland; by the time that it engaged with a peace-process alternative, it had become clear that their violence was not achieving this anticipated goal; and part of the thwarting of its violent activities arose because of intelligence-based state counter-terrorism operations.[61] The calculations of the PIRA and its political party Sinn Fein were made against a background of emerging political opportunity and ongoing belligerent stalemate; the latter helped to clarify the comparative advantages of pursuing the former.

What about other elements of counter-terrorism in Northern Ireland? Interim concessions in the form of temporary terrorist ceasefires did emerge at various points during the Troubles. The tactical issue of publicity was of huge importance. In the UK's own assessment of the campaign during the Troubles:

> For several reasons Information Operations were probably the most disappointing aspect of the campaign.... [T]hey were ill-coordinated with other government bodies; they were reactive; and often missed significant opportunities. The absence of a government information line was often exploited by the terrorist, sometimes with operational or strategic consequences. Constant criticism in the republican media, notably the *An Phoblacht* newspaper, was not seriously challenged by Government, NIO or Army Information Operations. Part of the reason for the ineffectiveness lay in the lack of a single unitary authority for the campaign, and the lack of a joint forum to agree Information Operations priorities, messages and means of dissemination.[62]

Atrocities by the PIRA and other Irish republican terrorist organizations did repeatedly give publicity gifts to the state[63] and (echoing experience elsewhere)[64] there was a paradox in terrorist publicity. Yes, terrorism gained headlines. But this was for acts which most people found repellent, especially if the victims were civilian. Yet it remains true that the state was not systematically effective in gaining a tactical-publicity victory during the Troubles, and there were far too many instances of state heavy-handedness or brutality for the picture to be a clear one. The January 1972 Bloody Sunday atrocity, in which British soldiers fatally shot fourteen unarmed civilians at an anti-internment march in Derry, lastingly undermined the UK state's record in Northern Ireland. Fifty years after the killings, even mainstream Irish voices remained outraged both by the events of that day and by the initial British attempt to cover it up. As an *Irish Times* editorial put it in 2022:

> as well as remembering the victims—some of them still children—whose lives were so cruelly cut off fifty years ago, we must also bear in mind the wider consequences of misgovernment. We must reflect on how democracies can fatally undermine themselves through the practice of impunity and the closing of official ranks.[65]

Moreover, there would have been no march against internment on 30 January 1972 had internment not been introduced one-sidedly, and unfairly in many cases. Indeed, the UK's own analysis of its military record during the Troubles is candid about the publicity effects of internment:

> Operation Demetrius, the introduction of internment, was in practice an operational level reverse. A considerable number of terrorist suspects were interned: the net total of active IRA terrorists still at large decreased by about

400 between July and December 1971. A very large amount of intelligence had been gained: the number of terrorists arrested doubled in six months. However, the information operations opportunity handed to the republican movement was enormous. Both the reintroduction of internment and the use of deep interrogation techniques had a major impact on popular opinion across Ireland, in Europe, and the US. Put simply, on balance and with the benefit of hindsight, it was a major mistake.[66]

The enduringly negative publicity secured by these episodes is clearly not its most important legacy. But it does represent a major aspect of UK counter-terrorism tactical work in Northern Ireland that proved counter-productive.

Treatment of prisoners often also generated negative publicity for the state. The 1976–81 prison war in Northern Ireland represented one major example. The republican prisoners who staged hunger strikes to campaign for political status during 1980–1 formed the tragic apotheosis of this struggle. While the second of these strikes ended without the prisoners securing their demands, considerable damage had been done to the UK's reputation and image,[67] and support for republicans themselves had been intensified in the North. A UK Northern Ireland Office Memo of August 1981 made clear the NIO view that the 'political and social effects' of the [Irish republican] hunger strike included 'a sharp increase in the influence of the Provisional IRA in Catholic areas, the alienation of much of the minority community in Northern Ireland from Government and the disruption of political life'; 'The Provos and INLA have gained a new batch of recruits. The feeling of alienation, bitterness, and frustration we detected in Catholic areas in May has grown steadily stronger. People are becoming anti-British and less ready to give the system their support.'[68]

Long-term historical realities were relevant here, and might (perhaps should?) have informed a more subtle policy on the part of the authorities. Painful hunger strikes had occurred during the Irish revolution of the early twentieth century,[69] and earlier in the modern-day Troubles there had also been republican adoption of the same tactic. In 1972, dozens of prisoners went on hunger strike in Belfast, Armagh, and Maze Prisons 'seeking political status'.[70] The emotive strength of this kind of development was reflected in messages such as that sent from County Fermanagh to Secretary of State William Whitelaw in June of that year: 'Are you going to let men and women in hunger strike die?'[71]

Of course, the issues involved in the prison protests were complex. The prisoners' central aim was to emphasize the political, rather than criminal, quality of their struggle and that of their republican comrades outside the jails: 'Attempts to criminalize us were designed to depoliticize the Irish national struggle.'[72] The state wanted to avoid a stress on the political dynamics of the terrorists' actions, fearing it might lend some legitimacy to them. In October 1980, the NIO presented a detailed account of its own perspective, giving background to the prison dispute, the introduction of special category status for prisoners in 1972, and its phasing out from 1976 onwards. The crucial and important context was ongoing, murderous violence.

> The government is seeking, in the face of a prolonged terrorist campaign in which more than 2,000 people have been killed and over 23,000 injured, to maintain the rule of law. It is an essential element of the government's approach that those found guilty after due process of law shall, if they are sent to prison by the courts, serve out their sentences in prison conditions which are as fair and humane as possible. Any such prison system must rest upon compliance with a set of rules which apply to all convicted prisoners, not just to some of them. The declared objective of the protesting prisoners and those who support them is to secure the restoration of a form of special treatment for certain offences. The prisoners concerned, however, are in no sense political prisoners detained for what they believe (no prisoner in Northern Ireland prisons is held there because of his [sic] political convictions); their claim to be 'political' stems from the supposed motives for their crimes. On examining these offences it can be seen that this claim is false; of the 346 prisoners taking part in the dirty protest on 24 September 1980, 54 had been convicted of murder, 33 attempted murder, 77 of firearms offences, and 102 of explosive offences.[73]

These are non-trivial points. The prison protests were indeed intended to legitimize the brutal violence that republicans practised beyond the jails. But the political motivation behind this callous terrorist violence was equally clear and, in terms of publicity, to deny this was self-damaging to the UK state. Arguably, recognition of political motivation need not justify or mitigate the illegitimacy of terrorist organizations' violence, and a subtler governmental response would probably have helped the cause of the British authorities. For, while the UK attempted to refer to the existence of international sympathy for its stance,[74] considerable damage was indeed done to the public perception of its cause by its seeming denial of the political quality of admittedly horrific violence. However explicable the authorities' instinct here, the public stance (such as that adopted at the first emergence of the 1980 hunger strike) was unlikely to avert the publicity damage eventually done.

Compounding these problems, the publicity emerging from UK military heavy-handedness in Northern Ireland reinforced the strength and resources of anti-state terrorist organizations.[75] Some counter-terrorism actions in particular caused adverse publicity. This heavy-handedness can be understood through the words of experienced British policeman John Stalker regarding his 'Shoot-to-Kill' inquiry of May 1984: 'I was asked to undertake an investigation in Northern Ireland that very soon pointed towards possible offences of murder and conspiracy to pervert the course of justice, these offences committed by members of the proud Royal Ulster Constabulary.... It cannot be disputed that in a five-week period in the mid-winter of 1982 six men were shot dead by a specialist squad of police officers in Northern Ireland. The circumstances of those shootings pointed to a police inclination, if not a policy, to shoot suspects dead without warning rather than to arrest them.'[76]

Allegations of brutality by state forces of course suited anti-state organizations and need to be assessed in that light. But the evidence of repeated mistreatment is too strong to be simply dismissed, and it included too many instances which gave advantage to opponents. The injuries inflicted in the Maze Prison at Long Kesh even before the 1976–81 prison protest offer a telling example. In September 1972, a group of Catholic priests highlighted the fact that fifty-three prisoners had sustained injuries (including bruising, dog bites, and wounds requiring stitches) at the hands of prison warders and British soldiers there.[77]

What of the tactical undermining of opponents? As noted, the state did manage to infiltrate groups such as the PIRA and the UDA and, on that basis, to forestall many planned terrorist operations. Such action also generated anxiety, mistrust, doubt, and even paranoia among the enemy. Morale could be damaged by the knowledge that your own ranks had been infiltrated by the state; in one ex-PIRA member's words, 'someone close to me who knew my intentions must have informed on me. This was a very bad feeling.'[78]

In terms of maintaining control over a population, the UK state largely succeeded, although some areas in the North were almost enemy zones at some points. But the challenge of population control did not only involve anti-state terrorism. The capacity of violent opponents (especially loyalist terrorist ones) to destroy the 1974 power-sharing government in Belfast represented a major failure by the UK state. The Ulster Workers' Council

(UWC) drew on support from loyalist paramilitaries to undermine the inter-communal Northern Ireland Executive through an aggressive strike in that year.[79] This represented a notable failure of state control.[80]

Organizational strengthening by the state did occur. It became increasingly difficult for the PIRA to kill members of the British Army; and the increased coordination of the different wings of the counter-terrorist elements of the state helped to account for the tactical operational successes discussed above. Of course, coordination in intelligence work was not always smooth. Whistleblower Martin Ingram (Ian Hurst) commented on the intelligence community and its work in Northern Ireland that, 'The internal squabbles could be childish at times. Without doubt they led to inefficiency, and sometimes to loss of life. ... The problem that the [intelligence] community suffered was a basic one—little or no coordination, due mainly to the intense distrust between the various agencies.'[81] A 1980 Report (commissioned by the UK and carried out by a future Director General of MI5, Patrick Walker) found that, 'The relationship between Special Branch [SB] and CID [Criminal Investigation Department] is close but sometimes uneasy. An element of competition can appear in their work in the areas where their functions overlap.' Walker recommended greater integration, less duplication, and a more dominant role for Special Branch in relation to intelligence work.[82] And the organizational situation did improve and gain strength. The Tasking and Coordination Group (TCG) was initially set up in Belfast in 1978, to coordinate, allocate, and monitor the activities of intelligence-gathering bodies (RUC SB, MI5, Army Intelligence). The same model was subsequently adopted elsewhere in Northern Ireland, and helped to produce more harmonious and integrated counter-terrorism work. Maurice Oldfield (1915–81), Chief of the Secret Intelligence Service (SIS) during 1973–8, became Security Coordinator in Northern Ireland during 1979–80 and helped to improve coordination between the RUC, MI5, MI6, and the British Army regarding intelligence work in the North. By the early 1990s, the coordination challenges between different wings of the state had been significantly addressed.

II

Beyond the strategic, the partial strategic, and the tactical levels systematically assessed above, counter-terrorism could also be judged to work if it offered its practitioners inherent rewards (as laid out in my introduction), perhaps autonomous of strategic or tactical outcomes but still beneficial for those who enjoyed them. How evident was this in the Northern Ireland Troubles? On the morning of 3 June 1991, in the County Tyrone village of Coagh, the SAS killed three PIRA men (Pete Ryan, Tony Doris, Lawrence McNally) while they were apparently on their way to try to kill a part-time UDR member. Some took satisfaction from the episode. Democratic Unionist Party (DUP) leader Ian Paisley responded by claiming that, 'The Army has once again demonstrated its ability to take out of circulation the IRA murdering thugs who are carrying out a campaign of blood in our province.'[83]

Rewards, such as they are, were complemented, of course, by some costs. Stella Rimington spent twenty-seven years in the UK Security Service (MI5), including time working on counter-terrorism, and her comments may reflect the experience of many:

> Counter-espionage work is not a glamorous business, however it has been presented by the spy-story writers. It is hard work. It is all about painstaking and rigorous analysis, the detailed following up of snippets of information and perseverance in the face of disappointment.... One of the problems of working in a secret organization is maintaining a normal life outside one's work. Making and keeping friends, even the sort of loose circles of friends and neighbours which most people have, is not straightforward when you are required to keep your employment secret.[84]

But inherent rewards there were. Regarding MI5, Rimington herself suggested that the motivation 'is not money. Unlike major companies in the commercial sector, MI5 does not set out to offer top quartile executive pay with bonuses and share options. Motivation is complex. It comes from a combination of the intrinsic interest and excitement of the work itself...and a sense of the importance of the job to be done. There is also a strong sense of loyalty to the organization, to colleagues, and to the country, however that is defined and however unfashionable that may sound.'[85] Close comradeship is evident from other counter-terrorism sources too. Martin Ingram (Ian Hurst) suggested of the FRU that, 'We were a small, tight-knit bunch in the FRU. We worked together, drank together, played football together. We were each other's support system.'[86] Among British soldiers fighting the PIRA, there could be very profound, emotionally intense, enduring comradeship, complemented by a sense of doing something important, fascinating, exciting, enjoyable, exhilarating, and ultimately life-saving.[87] For British agents within terrorist organizations, there were some financial rewards as well as a sense of doing good; but there were also very serious (at times, lethal) dangers.[88] One PIRA member who worked as an RUC Special Branch agent noted that he made money from the process, that he gained excitement and respect as well as danger through the work, but also that he had a desire 'to do everything possible to end the sectarian violence and save innocent people's lives'.[89]

The risks and dangers involved in counter-terrorism could affect many different kinds of actor. Between 1974 and 1993, twenty-nine Northern Ireland prison staff were murdered (almost always by republicans).[90] And one Northern Ireland prison officer, recalling the highly pressured context of the 1970s, observed that, 'I was often working a seven-day week and it was really like you were the one who was serving time'.[91] During the Troubles as a whole, the death toll among those fighting terrorism was a terrible one: between 1969 and 1998, 709 British soldiers and 303 RUC members died.[92]

III

Systematic analysis of the layered kinds of success involved in Northern Irish counter-terrorism therefore points to a partial strategic success for the state. This was grounded on a complex tactical picture, involving operations, interim concessions, publicity, the undermining of opponents, control over relevant populations, and the strengthening of the state. That tactical work was far from flawless. But it did, collectively, help to produce the circumstances within which a major peace process could emerge. Inherent rewards were complemented by some terrible costs paid by counter-terrorist personnel.

The lives saved by the Northern Ireland peace process should be remembered. One estimate is that, during the twenty years following the 1998 Belfast/ Good Friday Agreement, around 2,400 people in Northern Ireland were spared the violent death that might have been expected for them had that Agreement not provided the basis for an effective peace.[93] If one adds those who would also have been injured and traumatized had the conflict continued at previous levels, then the scale of that peace-process achievement becomes even more significant. To the extent that counter-terrorism formed a part of this process, it should be seen as having saved and protected many lives.

The historical approach adopted in this book emphasizes the role of contingency, and the capacity of Irish politics to embody such a theme has again been reflected in recent years. The 2016 UK referendum on European Union (EU) membership involved major developments outside Northern Ireland which profoundly affected the politics of that region. While the UK as a whole voted to leave the EU, a majority in Northern Ireland opted to stay. Moreover, and significantly, Irish nationalists in the North overwhelmingly voted Remain rather than Leave,[94] and the implications of Brexit (especially concerns about where the UK border with the EU would effectively lie) reinforced polarization in Northern Ireland, and between the UK and Irish governments.

At time of writing, there is no compelling evidence that the effects of Brexit will lead to a return to high levels of violence in Northern Ireland. Indeed, compared with enormous crises faced in the past, the North's problems in the post-2016 period might be judged extremely limited.

But several points do need to be registered as we conclude our assessment of UK counter-terrorism in relation to Northern Ireland.

First, Brexit did reflect an enduring problem in relation to the UK, in terms of British indifference to Northern Ireland other than during crises there. It is not the case that Brexit can be explained entirely on the basis of English nationalism; if those in Scotland, Wales, and Northern Ireland who voted in 2016 to Leave the EU had instead voted to Remain, then Brexit would have been decisively defeated.[95] Despite this, and in spite of what then Prime Minister David Cameron has subsequently implied,[96] it remains true that not enough attention was paid in Britain (and most importantly in England), during pre-Referendum politics, to Brexit's possible effects in and on Northern Ireland.[97]

Moreover, there is no doubt that Brexit has shifted opinion among Northern nationalists away from an endorsement of the UK as a long-term political home. Being dragged against their will out of the EU, on the basis of British voters' preferences, proved understandably unpopular among the Irish nationalist population in the North. In the years 2004–10, very clear majorities in Northern Ireland repeatedly favoured it remaining within the UK; even among the Catholic population in the North, most of these years saw only a minority of people saying that they favoured a united Ireland.[98] In the period after Brexit, there emerged something of a shift from this position, and a small number of surveys in Northern Ireland since 2017 have even suggested that support in the North for Irish unification was close to or at 50 per cent.[99]

This change should not be exaggerated. Those from a Protestant back-ground in Northern Ireland remain more likely to oppose Irish unity than those from a Catholic background are to support it,[100] and—on examination of the widest range of evidence—the immediate prospect of a majority in the North voting to leave the UK remains elusive. One reputable and recent poll (2022), found that only 30 per cent of people in Northern Ireland agreed with the statement 'I would vote for a united Ireland tomorrow', and only 33 per cent with the statement 'I would vote for a united Ireland in 15–20 years' time'.[101] Again, the 2021 Northern Ireland Life and Times Survey, asking 'If there was a referendum tomorrow, would you vote for N. Ireland to unify with the Republic of Ireland?', found that 34 per cent of respondents said yes, while 48 per cent said no.[102] There is also the issue of attitudes in the rest of Ireland, where rhetorical endorsement of the goal of Irish unity must be read through the lens of governmental caution about it, and also a frequent lack of popular prioritizing of the issue in political practice.[103]

In considering the legacies of Northern Ireland counter-terrorism, and the challenges facing it in the future, there exists therefore an ironic situation in which most people in the North continue to want to live in the UK, while most people in the UK have little interest in or commitment to Northern Ireland. The immediate and important challenge of preventing loyalist and republican terrorists is therefore overlain with the question of complex and unknown futures. If Ireland's long past is a crucial element in understanding counter-terrorism during the Northern Troubles, then respecting the complexities involved in long and contingent futures is equally important. Irish experience suggests that it would be unwise to try to predict with too much certainty how the Northern Ireland story will develop over future years. It remains the case that most people in the North do not now believe violence to be the best way of achieving political progress. Most people in both political communities during the conflict tended to support political parties that were not aligned with terrorist organizations.[104] Even more emphatically, the post-Troubles population of the North overwhelmingly rejects groups that endorse violent methods.[105]

But nobody can predict with any certainty how long the Irish border will persist. A borderless Ireland is far from inevitable, despite the growth during recent years of the influence of Sinn Fein across the island.[106] But it is certainly possible, and more so after Brexit, that a future border poll in Northern Ireland might yield a majority in favour of Irish unity.[107] Such an outcome would not be likely to involve the ending of political tensions in Ireland,[108] and indeed the possibility does exist of serious loyalist terrorism emerging at that stage to resist such a shift.

Peace, therefore, should not be taken for granted.[109] I myself think that there currently exists no likelihood of a return to large-scale violence. The issues of power and equality within Northern Ireland which lay at the root of nationalist disaffection do not exist as once they did; and the state no longer gives easy wins to terrorist entrepreneurs in the way that it tragically did in earlier phases of Northern Irish politics. So there do not currently seem to exist the conditions for any serious threat to the peace that was hard-won by so many people.

But the best way of avoiding that possible return would be to reflect on what went wrong and what went right in relation to counter-terrorism in the past. The greatest successes emerged when there was realism about the need to contain rather than entirely to remove terrorism from society; when there was attention to addressing the political root causes of which the violence was a symptom (and which the UK state needed to address); when there was restraint in the use of military methods to counter terrorism, rather than exaggeration of what such methods could achieve; when high-grade intelligence was used appropriately to undermine terrorist organizations and their activities; when the state adhered to the proper rule of law and the democratic oversight of those legal processes; when there was coordination rather than stove-piped disharmony between different wings of the UK state, and between the UK and other states; and when there was credibility in the claims, honesty, and arguments used in relation to trying to limit the violence.[110] When these principles were adhered to, UK counter-terrorism in Northern Ireland worked best; when they were ignored, terrorism was inflamed rather than constrained.

How far such insights will prevail is itself uncertain. Stella Rimington later fictionally combined Irish republican and global jihadist threats in a craftily plotted novel published in 2006. Entitled *Secret Asset*, the book is one in which the PIRA have placed a vengeance-seeking mole within MI5; amid the waning of Irish republican violence against the UK, this character instead aids jihadists in a plot within England.[111] This aptly reflects, perhaps, the transition of security focus that was under way during the late twentieth and early twenty-first centuries. Rimington's time as Security Service Director General (1992–6) coincided with a crucial period in the early, jagged process towards peace in Northern Ireland. By the time that *Secret Asset* appeared in

2006, the PIRA had formally brought their armed struggle to an end, and the 7/7 attacks of 2005 had painfully reinforced the tragic reality of the jihadist threat against the UK. Not everything that could have been learned from Northern Ireland about effective counter-terrorism was reflected in the UK's approach to the post-9/11 War on Terror. But, in Ulster as elsewhere, there is much that can honestly be derived from systematic reflection on what worked well, and on what did not, during the Northern Ireland Troubles.

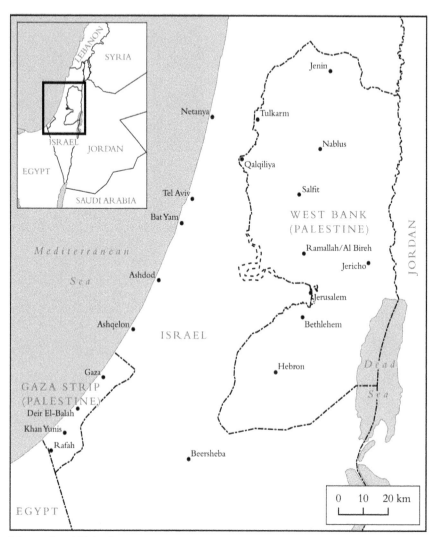

Map 3 Israel/Palestine

PART III

Israeli Counter-Terrorism

6

From the Six-Day War to the Gaza Tragedy

I

If Israel/Palestine provides an example of rival terrorisms,[1] then it also embodies rival nationalisms from which the terrorist violence has emerged. Part of this has involved Jewish and Muslim-Palestinian support for terrorism over many years. But there has also been a long tradition of people in each of these communities opposing the use of terrorist violence by their own side.[2] Scrutiny of Israeli counter-terrorism must therefore recognize the multi-sidedness of the region's violence, but also the multi-sidedness of attempts to limit it as well. Colum McCann's extraordinary novel *Apeirogon* is beauti-fully suggestive here, its empathy being comprehensive, and its title indicat-ing the importance of recognizing an infinite number of sides involved in Israel/Palestine and its peoples.[3]

As with other counter-terrorisms, the Israeli case cannot be understood without historical reflection and context. Religious and political conflict for power over the territory now comprising Israel/Palestine is centuries old.[4] The legacies of imperialism and colonialism in the Middle East are important contexts for the current conflict. The First World War and its complex aftermath were crucial to the establishment of Middle Eastern boundaries and states.[5] And if nationalism is understood as involving the interweaving of the politics of community, struggle, and power, then the rivalries and conflicts that have generated so much terrorism in the region become normalized within experience we can see around the world (attachment to the particularities of national territory, people, descent, culture, history, ethics, and exclusivism is common for most people in the modern period).[6] In Israel/Palestine it has not been possible to resolve the tensions generated by competing claims to legitimacy and authority in relation to these themes, and this provides the foundation on which terrorism and counter-terrorism have been built.

For, whatever the ancientness of rival Jewish and Palestinian claims, terrorism and counter-terrorism in Israel/Palestine have centrally emerged over questions of state power. Terrorism has existed in relation to the state of Israel throughout its history: terrorism originating from those fighting to make the state exist, and terrorism subsequently emerging as a means of fighting to eradicate the state or to defend it. Israel had been established in the 1940s partly as the consequence of a Jewish terrorist campaign that had been aimed at establishing such a state, so there has existed a measure of ambivalence on the part of this state (as of so many states) in regard to terrorism as such. No such ambivalence, however, has been shown by the Israeli state towards Palestinian terrorists. Many of the latter have been driven by a desire to destroy the state of Israel itself, and they have produced very high levels of attacks relative to the size of the population concerned.

In Israel, therefore, counter-terrorism has ultimately been about rival nationalisms, and the violence involved in their pursuit and defence. Terrorism and responses to it have been important aspects of the experience of the region for over a hundred years, and the pre-history of even that period forms part of the necessary context too. In 1882, there commenced the first Zionist immigration into Palestine (then part of the Ottoman Empire). Muslims and Christians already inhabited the region at that point, but part of the migration process involved significant numbers of East European and Russian Jews seeking to escape oppression during the late nineteenth century. By 1914, the population of Palestine was in the region of 722,000 people: approximately 582,000 Muslim Arabs, 80,000 Christians, and 60,000 Jews (about 8 per cent of the population).[7]

The political situation was transformed when, during World War I, in 1917, the UK conquered Palestine, taking the territory from the Ottoman Turks. On 2 November 1917, the Balfour Declaration expressed UK support for the establishment of a national home for the Jewish people in Palestine. In the short term, part of the aim here was to rally Jewish support (especially in the US) for the ongoing war effort. The long-term effects of the Declaration were profound. Arabs already living in Palestine understandably felt angered at this development. Some British officials had already offered similar-sounding promises to Arab leaders. The UK therefore possessed a double commitment, to groups whose interests and demands clashed directly with one another. The importance of this was reinforced when, on 22 July 1922, the League of Nations formally granted the Mandate for Palestine to the UK, giving authority for Britain to administer the region. During the 1920s there was violence in the region between Jews and Arabs, each seeing the other as a threat.

As with this post-First World War legacy, so too global events during the mid-century were crucially significant. The 1930s rise of Hitler, and the German persecution of Jews, led to further Jewish emigration to Palestine. During that decade also, organized Palestinian resistance to British rule and against Jews reflected a desire to end Jewish immigration, to limit the sale of land to Jews, and to establish a Palestinian national government.

The 1936–9 Palestine Arab Rebellion, or Arab Revolt, involved the targeting of Jews, but also some Arabs and some British victims too. The causes behind this first major Palestinian uprising were primarily twofold: the economic effects of the 1930s depression, which caused considerable hardship; and the ongoing growth in European Jewish immigration into Palestine, which was opposed by non-Jewish Palestinians. Palestinian nationalism was strongly evident here,[8] and the politics of community, struggle, and power again clearly represent the appropriate explanatory framework for the contemporary violence and tension.[9]

The dynamics of inter-war violence in Palestine were therefore fundamentally tripartite, involving the British, the Jewish Zionists, and the Palestinian Arabs. Empire, decolonization, global conflict, and competing nationalisms were all important. During the 1940s, Jewish terrorist organizations—the Irgun Zvai Le'umi (National Military Organization) and the Lohamei Herut Yisrael (Fighters for the Freedom of Israel, also known as Lehi and as the Stern Gang)—violently challenged British rule in Palestine.[10] When in 1947 the UK decided to leave Palestine, events in the region were again vitally affected by global dynamics, especially the legacy of the Second World War and the Holocaust. On 29 November 1947, the United Nations approved a resolution to partition Palestine into two states, one Jewish and one Arab. Palestinian Arabs opposed this; a subsequent war saw Israel take control of most of Palestine. Israel declared its independence on 14 May 1948, an event remembered by Palestinians as Al Nakba, the catastrophe, or the disaster. The forced displacement of thousands of Palestinians during the 1948 war (over 700,000 Palestinians were forced into exile) represented a tragedy with long-lasting consequences. For their part, Jews had experienced the atrocity of the Holocaust, or Shoah: state terrorism by the Nazis on an appalling scale.

British disengagement from Palestine was not just due to terrorism, and the UK response to Palestine involved more than a counter-terrorist dimension. The decline of empire, the existence of economic pressures and exhaustion, strategic commitments and considerations elsewhere, the question of public opinion in Britain, the understandable sympathy felt towards the establishment of a Jewish state after the Nazi Holocaust—all of these played their part in leading to the decision to leave Palestine. But terrorism did also play a part in that decision and there had therefore (before the establishment of Israel) been a counter-terrorist aspect to politics and government, given the levels of violence. In 1938 alone, there were 5,708 recorded terrorist incidents in Palestine.[11]

The 1948 declaration of the state of Israel therefore occurred not only against the backdrop of two national tragedies and catastrophes, but also on the foundation of terrorism and counter-terrorism already being part of the fabric of regional politics. Upon its establishment, Israel also immediately faced a coalition of Arab states. On 15 May 1948, British troops withdrew. Arab armies from neighbouring territories invaded; but when the fighting ended in February 1949, around three-quarters of the Mandate territory was occupied by Israel.

Israel had therefore been born amid violence between Jews and Palestinians, against the context of huge communal tragedies, and with a war between Israel and its neighbours (Syria, Jordan, Egypt, Iraq, and Lebanon). The governments of young states typically and explicably focus much attention on protecting their sovereignty[12] and Israel did face genuine and violent threats, internally and externally, from its birth. Indeed, the establishment of the state of Israel had changed the nature of the violence and tensions that obtained in the region. The continued existence (or not) of this state, and its relations with neighbouring states, now became enduringly crucial issues. Wars were to occur between Israel and its Arab neighbours (in 1948, in 1967, in 1973). And, as we have seen, by the time Israel came into existence there had already developed a pattern of cyclical, mutually stimulated violence between Jews and Muslims in Palestine, affected significantly by the contingent role of external actors and contexts. This pattern echoed and re-echoed. Retaliation, revenge, and tragically vengeful reprisal (so bloodily persistent a set of themes in terrorism and counter-terrorism)[13] again and again arose in Israel/Palestine. Establishing a secure existence formed a crucial aspect of the Israeli state's approach towards its violent enemies. Israeli counter-terrorism sought to signal resolve, and clarify the state's capacity to inflict heavy costs upon enemies.

Counter-terrorism and terrorism shape each other,[14] and the early decades of Israel's life saw the emergence of major terrorist opponents. The Palestine National Liberation Movement (Fatah) was founded by Yasser Arafat and Khalil al-Wazir in 1958; Assifa, Fatah's armed wing, was to launch many terrorist attacks in Israel (Fatah's first terrorist attack was carried out in December 1964).[15] Together with other Palestinian groups, Fatah came under the flag of the Palestine Liberation Organization (PLO), and grew to be dominant within that group. Established in 1964, the PLO itself was a nationalist Palestinian movement, deeply opposed to Israel and seeking to replace it with Arab statehood over the Mandate territory. Again, and infamously, terrorism was to be one of this organization's enduring methods.

This brief recap of long-rooted enmities demonstrates some of the inheritances shaping the realities of modern-day Israeli counter-terrorism. Nationalist rivalries over legitimate statehood; long-rooted local cycles of brutal violence; the establishment and sustenance of major non-state terrorist groups; tense animosity between mutually aggressive states within the region; the sometime intensification of local violence generated by global politics and events—all of these factors made the addressing of terrorist threats to Israel both more important and more difficult.

II

During the period 1967–87, Palestinian nationalist terrorism became and remained prominent. It took many different forms over these years, including hijackings, hostage-taking, suicide attacks, and lone-actor assaults using crude weapons such as knives. It occurred both within Israel/Palestine and in other parts of the world against Jewish targets.

In 1967 itself there occurred an epoch-defining episode in the Six-Day War: a decisive victory for Israel over its Arab neighbours, a devastating defeat for the Arab armies, and a brief conflict which left Israel in possession of all the territory that had comprised Mandate Palestine. Israel now controlled the Sinai Peninsula, the Golan Heights, the Gaza Strip, the West Bank,[16] and East Jerusalem. The border change of 1967 was to prove crucial. Pre-1967, the West Bank had been controlled by Jordan, and Gaza by

Egypt. Now, after the Arab states' humiliating defeat, Israel ruled over approximately 1 million Palestinian Arabs, the latter not recognizing Israel as possessing legitimate authority over them.

Those who condemn the long-enduring Israeli occupation of these territories frequently have very strong arguments to make. But it is vital also to understand (not least if one does seek the ending of the occupation) the precise reasons for this 1967 conflict, and this is less frequently evident in writings and commentary on the subject. It was not an inevitable war. Professor Chris Blattman's superb study of the reasons behind violent conflict offers a valuable lens through which to read the episode. Blattman outlines five main (interlocking) reasons that help explain why people turn to war. There can be 'unchecked interests', where societies have too few 'checks and balances' upon power: 'when the people who decide on war aren't accountable to the others in their group, they can ignore some of the costs and agony of fighting.' There are 'intangible incentives', where violence can secure for you something of high value such as dominance, revenge, dignity, status, ideological rewards, or 'pleasure in righteous action'. There is the role played by 'uncertainty', where you do not possess the information to clarify things like your opponent's intentions or strengths. There can be a 'commitment problem': if your rival is getting stronger, how can you be sure of their commitment not to use that growing capacity against you in the future? Might it make better sense not to trust such a future commitment, and instead to attack them now? Finally, there are 'misperceptions', when we think ourselves stronger than we are, exaggerate the likelihood of our victory, and misattribute ideas and qualities to opponents whom we sometimes stereotype and dehumanize; all of these misconstruals make conflict more likely.[17]

All five factors are clear and significant in the complex and contingent build-up to the 1967 war as follows.

Unchecked interests: Egyptian leader Gamal Abdel Nasser was indeed a largely unchecked ruler, something which played an important role in generating the war. Intangible incentives: Israeli politicians were motivated by a profound ideological and emotional commitment to preserving their precious state, and many Israeli people wanted revenge for the repeated attacks upon their own communities by Palestinian and Arab forces; for their part, some anti-Israelis passionately wanted to inflict righteous vengeance and destruction on Israel, while others (like Nasser) felt a need to replace humiliation with proud dignity and honour. Uncertainty: there was too little precise knowledge about others' intentions, and this intensified anxiety on all sides about the possibility of imminent attack against them. Commitment problems: Israeli decision-makers lacked faith that their Arab and Palestinian neighbours could be trusted not to use opportunities to destroy Israel in the future; therefore there seemed value in a strike at this particular moment. Misperceptions: some opponents of Israel profoundly exaggerated the likelihood of their being able to destroy the Jewish state through conflict, just as some Israeli politicians inflated the dangers that in fact existed for Israel.[18] If we want to unravel the long legacy of the 1967 war, then appreciation of these complex, contemporary dynamics behind it are essential.

In addition, other conflicts erupted episodically. In 1973, the Yom Kippur War further reflected Israeli strength, when a surprise attack by Egypt was followed by Israel again defeating Arab armies. Against this context of expanding Israeli power and authority, Palestinian violence (in response and resistance) also expanded. Between 1971 and 1982, Palestinian attacks in Israel itself killed more than 200 civilians and injured over 1,500.[19] There were also increasing numbers of Palestinian terrorist attacks on Israeli targets outside Israel during the late 1960s and early 1970s. Most famous among these was the Munich Olympics outrage of September 1972, when Palestinian terrorists attacked Israeli athletes, demanding that Palestinian and some other terrorist prisoners be released in exchange for hostages. A clumsy rescue attempt left numerous people dead, including several of the terrorists, as well as the Israeli hostages and one West German police officer. Israel had, prior to Munich, deployed targeted attacks against some Palestinian enemies. But Munich intensified the sense that this approach was appropriate and necessary, so Israel focused attention on people involved in the 1972 atrocity, wherever they were in the world; Operation Wrath of God duly targeted numerous Palestinian terrorists and activists.

While Munich remains the most famous of all Palestinian terrorist operations, many other instances of anti-Israeli violence emerged during this period, hijackings conspicuous among them. The politically motivated hijacking of aeroplanes long predates late twentieth-century Palestinian terrorism.[20] But it was with Palestinian terrorist deployment of the tactic that hijacking became globally salient in a new way. Targeting Israel's national airline El Al, the Popular Front for the Liberation of Palestine (PFLP) hijacked planes and held their crew and passengers hostage with a view to securing the release of prisoners held by Israel. When security was improved for El Al flights, the PFLP and other groups then widened their targets. High levels of media attention resulted. But the Israeli tactical-operational response did make it a more difficult mechanism for terrorists to deploy. It is also true that plane hijackings lost momentum partly because of the negative response to the violence used,[21] an example again of the paradox of terrorist publicity: yes, terrorism gains headlines; but it does so for actions which most observers find repellent, especially if the victims are civilian.[22]

A certain pragmatism emerged in Israeli tactical-operational reactions here. The Rabin Doctrine[23] regarding responses to terrorist hijackings stipulated that if a military rescue operation were feasible then this should be the preferred option, but that if such an operation were not possible then negotiations with the terrorists could be pursued to try to release the hostages. The dramatic Entebbe raid of July 1976 reflected the opportunities, but also the dangers, involved in such an approach. Following the hijacking of a French aeroplane by a splinter group of the PFLP, the plane's crew and over 200 passengers (most of the latter being Jewish) were taken by the terrorists to Entebbe airport in Uganda, where they were held hostage. Some of the hostages were released. On 3 July 1976, almost all of the remaining captives were freed during a risky and very impressive Israeli military rescue operation, which also saw the terrorists killed.[24] The episode showed the lengths (including considerable physical distance, as well as elaborate tactical-operational activity) to which Israel was prepared to go to combat terrorism.

The dynamics between state and non-state actors here were fluid. From the mid-1970s onwards, Yasser Arafat led the PLO to combine violent tactics with an apparent openness to negotiation and diplomacy. Not unrelatedly, in July 1974 the PLO leadership opted to desist from terrorist operations against Israeli and Jewish targets outside Israel, having recognized that operations of this kind in fact damaged their diplomatic progress. The role of Arafat (1929–2004) was vital here. Elected PLO chairman in 1969, he led the movement for twenty-five years and became the most iconic Palestinian leader during a phase when Palestinian terrorism and the Palestinian cause emerged into global prominence. Fatah were to suffer after Arafat's death, his prestige having been crucial to them. By then, the PLO's approach had changed significantly, so much so that in December 1988 the organization announced that it now recognized Israel's right to exist, and indeed that it rejected terrorism.

This partial strategic success for Israel partly emerged because of intense, committed, and skilful counter-terrorist tactics by that state. Palestinian hijackers' frequent demand for the release of Palestinian prisoners in exchange for hostages reminds us that Israel had arrested and detained large numbers of Palestinians. And Israel had also been developing complementary political initiatives. With the 1978 Camp David Accords, relations between Israel and Egypt were regularized, and the latter regained the Israeli-occupied Sinai Peninsula.

More aggressively, Israel's June 1982 invasion of Lebanon (Operation Peace for Galilee) saw the crushing of the PLO in that country, where they then had their Headquarters. This assault, however, involved a combination of tactical success and strategic frustration. The invasion considerably damaged the PLO's military capacity. But it also had longer-term counter-productive effects for Israel. Instead of speedily withdrawing after their military success, Israel remained as an occupying force, and this generated the resentment which helped to create Hezbollah as a major anti-Israeli resistance movement.[25]

In the years 1967–87, therefore, we can read some of the complexities of Israeli counter-terrorism. Strategic success proved elusive, with the ongoing threat of terrorism continuing to represent a major challenge for the Israeli state. But partial strategic progress had been made, not least with the PLO being moved towards a position in which they would soon recognize Israel's right to exist, and turn away from terrorist tactics. There had been some impressive successes, whether in wars against neighbouring states in 1967 and 1973, effective operations such as that in Entebbe in 1976, or in the undermining of opponents through operations such as the Lebanese venture in 1982. Our framework helps illuminate some of the complications involved here, however. The tactical-operational success of the 1967 conflict generated the basis for long-term Palestinian resistance in the now-occupied territory of the West Bank and Gaza. The undermining of the PLO in 1982 in Lebanon was offset by the counter-productive creation of one of Israel's main enemies of future decades, Hezbollah. One of the painfully significant themes within Israeli counter-terrorism—the tension between operational brilliance and strategic self-harm—was already evident in this blood-stained period.

III

Similar shapes can be detected in Israeli counter-terrorism from 1987 onwards. The First Intifada (1987–93) began with the eruption of mostly non-violent Palestinian resistance to Israel. But state responses and subsequent interaction helped make this an episode of profound significance in relation to Israeli counter-terrorism.

In December 1987, demonstrations and riots emerged in Gaza and then in the West Bank. The reasons behind the outbreak are complex. Unemployment and economic hardship in the West Bank and Gaza had been accentuated in many Palestinians' minds by an appreciation of how much better daily life was for many Israelis. A sense of disappointment with what the PLO were seemingly able to achieve for people in the Occupied Territories was becoming especially felt by a younger generation, and so there was something of a generational dynamic to the uprising also. The existing seeds of a more fundamentalist version of Islamic Palestinian politics were also of relevance. The political background here was vital. Economic conditions in Gaza and West Bank were interwoven with Palestinian nationalist grievance. The 650,000 inhabitants of the Gaza Strip, the 900,000 inhabitants of the West Bank, and the 130,000 inhabitants of East Jerusalem mostly wanted to live in a Palestinian state, rather than as stateless people under what they saw as a harsh military occupation by Israel.

The specific trigger for the Intifada occurred on 8 December 1987, with an accident in Gaza. An Israeli truck hit a Palestinian car, and four Palestinians died as a result. Rumours spread that this was a deliberate act. Rioting and disorder began on the following day, spread throughout much of the Occupied Territories, and came to represent a major problem for the Israeli authorities. Though largely non-violent, the disturbances did include the throwing of stones and the use of Molotov cocktails. Yet it is hard to see the response of the Israeli authorities as anything other than heavy-handed and disproportionate. The Israel Defence Forces (IDF) reacted to the First Intifada in ways that reinforced Palestinian anger, playing into the hands of those Palestinians who sought a more aggressive stance towards Israel. The IDF's initial approach was to try to bring the Intifada to an end through military means. This was intended as something which would both bring calm and send a message to future would-be rioters that there would be high costs to such actions.

Whatever the intended deterrent effect, the reality was less than eirenic. The centre of gravity of the Intifada (literally, a 'shaking off': a shaking off of the occupation) lay in protests, strikes, civil disobedience, marches, and sit-ins. There was a withdrawal of work from Israeli farms and factories, and a withholding of taxes. Such mostly non-violent action led to wider participation, including the involvement of large numbers of young people. Indeed, despite the headline-seizing quality of Palestinian terrorism, Palestinians have proved far more likely to participate in civil resistance than in violence.[26] But as Israeli measures increasingly involved curfews, detention, house demolition, and deportation, Palestinian anger grew.

The episode changed the dynamics of resistance in other ways too. The 1987 Intifada located the struggle very much in Palestine, in contrast to the PLO who had had their Headquarters elsewhere. And the uprising gained international publicity and much sympathy for Palestinians and their aims. The excessive response from Israel damaged the country's image abroad, as the punitive measures were visible worldwide through international media. Various kinds of violence during the First Intifada claimed Palestinian lives. Between December 1987 and December 1993, Israeli security forces killed 1,095 Palestinians in the Occupied Territories; between December 1987 and December 1991, another forty-eight Palestinians were killed by Israeli civilians; and during the first three years of the Intifada, 359 Palestinians (mostly suspected of collaborating with Israel) were killed by other Palestinians.[27] Alongside this was the issue of imprisonment. During the 1980s, an average of 4,500 Palestinians were held in Israeli custody on any given day; during the First Intifada itself (1987–93), between 20,000 and 25,000 Palestinians were arrested each year.[28]

Out of this, and emerging partly from frustration at the ineffectiveness of the PLO, there grew a powerful new force for Palestinian resistance, and one whose terrorism would prove a central issue for Israel. Officially founded in December 1987, Hamas had emerged out of the Muslim Brotherhood in Palestine and embodied a grievance-driven Islamic nationalism with a very violent edge. Offering militant resistance to Israeli occupation, Hamas opposed compromise with the Israelis, and sought the destruction of Israel and its replacement by an Islamic, Palestinian state comprising the land covered by the British Mandate (Israel, Gaza, the West Bank). One of the organization's leaders expressed it crisply in 1995: Hamas aimed at 'the establishment of an Islamic state instead of Israel. . . . We will never recognize Israel.'[29]

Hamas therefore, in contrast to the contemporary PLO, explicitly stated a desire to annihilate Israel. It was also more emphatically Islamic in its politics than the PLO had been. By the spring of 1988 Hamas represented a major part of the Intifada in the West Bank and especially in Gaza; in March of that year Hamas carried out its first military action, an ambush in Gaza in which an Israeli water engineer was injured. As so often with terrorism, therefore, a mixture of longstanding and more recent communal grievances, together with a contingent trigger to action, brought about a revolt. The state's response to a largely peaceful movement had proved counter-productively harsh, and seemed to justify those people in the protesting community who advocated less compromising and more violent politics. It is a familiar (though contingent rather than inevitable) pattern across the long and interwoven history of terrorism and counter-terrorism.[30]

Support for Hamas grew during the 1990s, as many Palestinians saw their hopes for statehood diminish and their land being given to Jewish settlers. Part of this involved Hamas rivalry with the PLO, and the divergent approaches adopted by the groups towards peace with Israel. The First Intifada had seen strong divergence between the PLO and Hamas, the former engaging more and more in talks with Israel, and the latter challenging such an approach and instead adopting a more aggressive attitude. Relationships between terrorist groups and their intra-communal rivals often shape their actions, their capacities, and their fluidity over time. State responses are therefore affected and partly conditioned by the need to be attentive not just to the state/non-state relationship, but also to relationships and power struggles within the community of rebellion. Here, Israel's challenge was to try to address Palestinian terrorism while those who had practised it were increasingly divided in their approach, and with Hamas positioning itself as the opponent of the 1990s peace process for which the PLO was now the enthusiast.

Hamas increasingly became Israel's main enemy. Sympathetic states (including Iran) and a supportive Palestinian diaspora provided financial contributions. And the combination of lethal terrorism with a broader politics of community social service provision made Hamas a formidable opponent, and indeed a social movement[31] as well as a terroristic one. During the 1990s the organization came to dominance as the major force offering Palestinian resistance to Israel, overtaking Fatah (the largest faction of the PLO) and others in significance.

Hamas's terrorism, much of it aimed at undermining the Israeli/Palestinian peace process, both seized international attention and also generated terrible human suffering. The first Palestinian suicide attack in Israel occurred on 16 April 1993 and was carried out by Hamas. Suicide bombing had been deployed in other conflicts, but became a frequent part of Palestinian struggle from the 1990s on: there were 217 suicide bombings between

1993 and 2006 in Israel and the occupied Palestinian territories.[32] The tactic is hard to guard against (no state can search everyone or entirely prevent movement of population, and post-attack interrogation is impossible if the assault succeeds); suicide attacks also allow for their practitioners to target precisely, and there is no intention of getting the attacker out alive.

Explaining Hamas's use of suicide attacks must involve recognition of the layered complexity of motivation and engagement involved.[33] The organization considered the method effective, and intended that these gruesome attacks should further the organization's strategic goals of undermining, and ultimately destroying, the Israeli state. The annual number of Hamas suicide attacks during 1993–2006 demonstrates the nature of the problem that Israeli counter-terrorism had to face: 1993 (5); 1994 (4); 1995 (3); 1996 (3); 1997 (3); 1998 (1); 2000 (1); 2001 (19); 2002 (12); 2003 (11); 2004 (5); 2005 (1); 2006 (1).[34]

The headline-seizing nature of these brutal attacks somewhat eclipsed the other work that Hamas was increasingly doing. Politically, Hamas increasingly engaged in electoral and governmental work and, beyond this, the movement became extensively committed to social service provision for its Palestinian people. They have run hospitals, libraries, schools, mosques, and various kinds of social services centres.[35] For Israel, Hamas is a terrorist organization and understandably seen as such. For its own supporters, however, Hamas can be read through the lenses of the Israeli occupation under which they themselves have lived: Hamas was resisting an occupying power; it was offering political hope and authority; it expressed a religiously informed nationalism of high authenticity; and it daily addressed some of the urgent social needs of a disempowered people.

In the decades since Hamas emerged in the 1980s, the group has never seemed near to realizing its central strategic aim of destroying Israel. But nor have its efforts been trivial. It has eaten away at Israeli desire to control Gaza and the West Bank; it has sustained resistance against Israeli occupation; it has inflicted many vengeful attacks on Jewish enemies; it has helped undermine a peace process which it considered too compromising; and it has secured much publicity for the Palestinian cause.[36] Despite the non-violent activity practised by Hamas, there long remained a terroristic problem for the Israeli state to respond to, and it was reasonable to see the political, the social, and the terroristic parts of Hamas's work as interwoven. What was Israel's response to this terrorism?

IV

Strategic or partial strategic success for Israel would see the removal or very significant erosion of terrorist violence, and would therefore require some form of political settlement between Israel and the Palestinians. In 1991, meetings in Madrid between Palestinians and Israelis marked a significant indication that some on both sides indeed sought a strategic resolution rather than an ongoing conflict. The gatherings did not produce agreement, but this November 1991 Madrid Conference (supported by the US and by Russia) included participants from Israel, Syria, Lebanon, Egypt, and a joint Jordanian/Palestinian delegation. The multilateral discussions were focused on the possibility of Palestinian self-government, and the involvement of the US was crucial. Indeed, Israeli counter-terrorism in recent decades cannot be properly analysed without reference to both the regional context and also the vital support offered to Israel by its American ally.

The US has long had an engagement in Israeli and wider Middle Eastern politics. This has involved strong sympathy towards Israel and also repeated efforts to secure peace in the Israeli-Palestinian conflict, including repeated presidential involvement.[37] The Israeli/Palestinian conflict has possessed significance for America partly because of the economic importance of that Middle East region, alongside commitment by the USA to the idea that Israeli/Palestinian peace would be in the US national interest.

The 1990s initially seemed to offer possibilities for this kind of strategic or partial strategic success in undermining the basis for terrorist violence. In 1992, talks in Oslo began as a meeting sponsored by Norway. When Israeli Prime Minister Yitzhak Rabin approved the participation of official Israeli representation, these talks became the main way of Palestinians and Israelis discussing their relationship and a possible deal. With direct PLO involvement, things initially looked propitious, and the Oslo Accords of 1993 seemed to fulfil that great promise. The Oslo Declaration of Principles on Interim Self-Government Arrangements embodied a very substantial attempt to end the terroristic violence in the Israel/Palestine conflict. The Oslo Accords (as the agreement became known) provided for the setting up of a Palestinian Authority (PA), a semi-autonomous governing body in parts of the Occupied Territories (the West Bank and Gaza Strip). It was therefore an extraordinary moment when—on 13 September 1993, at the White House in Washington, DC, with President Bill Clinton presiding—Israeli

Prime Minister Yitzhak Rabin and PLO Chairman Yasser Arafat together signed the Declaration of Principles. With Gaza and the West Bank moving from Israeli rule to the authority of the Palestinian Authority and security forces, the deal effectively involved the normalization of Fatah and its transition from terrorist movement to political party. In 1994, Palestinian autonomy was established in these parts of the Occupied Territories, Israel withdrew, and power was ceded to the PLO-dominated PA. The Oslo Accords also saw the two sides recognize the existence of each other, and the PLO more effectively renounce terror. Given the brutal conflict being addressed, the language and commitments were very striking indeed:

> The government of the state of Israel and the PLO team . . . agree that it is time to put an end to decades of confrontation and conflict, recognize their mutual legitimate and political rights, and strive to live in peaceful coexistence and mutual dignity and security and achieve a just, lasting, and comprehensive peace settlement and historic reconciliation through the agreed political process.[38]

Peace processes can be acts of counter-terrorism. As historian Joseph Morrison Skelly has pointed out, despite their frequent protestations to the contrary, governments and their representatives have repeatedly engaged in diplomacy with terrorist groups. And the record of such endeavours has been very mixed. As Skelly's assessment shows, there can be benefits, costs, and limitations for those who choose the route of negotiation with terrorists.[39] But, whatever the broader pattern, and despite the ultimate frustration of peace-process expectation in Israel/Palestine, the emergence of Palestinian self-rule in Gaza and the West Bank did represent a moment of publicly witnessed hope for many people. Even when other groups (such as Hamas) persisted with their violence, Israel and the PLO continued to try to work together. Yitzhak Rabin had been in the IDF during the 1948 Arab-Israeli War, and he was IDF Chief of Staff in the 1967 war which had seen Israel actually take over Gaza and the West Bank; Yasser Arafat was the most iconic Palestinian resistance leader, Israel's most famous terrorist adversary. The symbolism of this deal was therefore extraordinarily powerful, both for those who supported it and for those whom it enraged.

Could Oslo secure strategic success for Israel in its desire to end Palestinian terrorism? Following the deal, some ambiguity remained. Yasser Arafat did indeed denounce terrorism and he acted against those practising it from the Palestinian side. But many in Israel doubted whether this latter work was done as committedly as it could have been and also considered Arafat to be using ongoing violence as a way of trying to secure further political concessions from Israel.

Violence did indeed continue. For their part, Hamas saw the Oslo agreement as a breach of faith with the Palestinian cause (which they interpreted in line with more maximalist objectives), and the organization considered Arafat and the PLO as having betrayed their people. Hamas therefore pursued military operations with a view to ending what they saw as an unacceptable and overly compromising peace process. In the period after Oslo, Fatah overwhelmingly desisted from terrorism; Hamas emphatically did not. As so often with peace deals, the implementation was even more difficult than the drafting and reaching of the agreement. The Israeli government saw the essence of the peace process as being that it would guarantee security, and that it would end Palestinian terrorism. Given this, it was very difficult for that government to pursue the process amid ongoing Palestinian violence.

On the other side, some Palestinians felt that the Israeli government was not keen to pursue the peace deal's commitments, and that Israel used Palestinian terrorism as an excuse for not doing so. The Israelis continued with Jewish settlement expansion within the Occupied Territories even during the Oslo period, thereby making many Palestinians doubt their good faith, or how much genuine potential there was in the process. Moreover, terrorism against the Oslo deal came from some on both sides: Palestinian terrorists but also Jewish extremists. Despite this, much support for the peace process endured. In September 1997, it remained true that, according to one credible opinion poll, more Israelis favoured continuing with the Oslo peace process than opposed it (47 per cent in favour; 33 per cent against).[40]

For Oslo to succeed in terms of strategic or partial strategic success in limiting terrorism, it needed to be part of a longer process. In September 1995, the Oslo II Accords expanded Palestinian autonomy. Shortly after this, however, Yitzhak Rabin was killed. Jewish terrorism had predated the establishment of the state of Israel, and it continued during the life of that state, with some activists rejecting the politics of compromise. The right-wing terrorist Jewish Underground, for example, had been formed in 1979, and engaged in revenge attacks upon Palestinians, in response to Palestinian assaults on Jewish settlers. During the 1990s, some horrific attacks occurred. On 25 February 1994, Baruch Goldstein killed twenty-nine Palestinians and injured over one hundred others in a Hebron mosque. Hamas, similarly hostile to the peace process, responded to Goldstein's appalling killings with a suicide attack. The peace process was thus assaulted violently from both sides, and Rabin had faced genuine pressure originating both from Jewish and from Palestinian sources.

Then, on 4 November 1995, Yigal Amir (a right-wing Jewish extremist, angered by what he saw as Rabin's treacherous betrayal of Israel in the peace process, and by the giving of Jewish land to Palestinians) shot and killed the Israeli Prime Minister. In doing so, along with Palestinian rejectionist terrorism, he helped also to kill the hopes held by so many, and to damage the 1990s peace process.

The Israeli response to terrorism had not only involved political engagement with this process. There had been more muscular aspects too, and a keenness to display a long-term resolve to fight terrorism. Such tactics predated the early 2000s but were intensified during those years. In July 2000, with the support of US President Clinton, a Camp David summit was held with the aim of producing a more final resolution of the Israeli/Palestinian conflict. Under the proposed arrangement, Arafat was offered Palestinian control over almost the whole of the West Bank and Gaza, as well as the handing over of Arab areas in Jerusalem. He refused the proposed deal, a rejection which strengthened the sense among many Israelis that Arafat had not been genuine about peace and that the Israeli government's engagement with the peace process had been unwise.

On 25 July 2000, the Camp David negotiations broke down. Before the end of the year, a Second Intifada (much more violent than the first) had begun. On 28 September, shortly after the return of Israeli Prime Minister Ehud Barak and his team from Camp David, Israeli opposition leader Ariel Sharon visited the Temple Mount/Haram al-Sharif in Jerusalem (accompanied by more than 1,000 Israeli police officers). Already a tense time, given the failure of Camp David, this was without question a provocative gesture since Muslims as well as Jews consider this a sacred site. Palestinians reacted angrily to Sharon's visit (though some have questioned how spontaneous those protests were).[41] Hundreds of Palestinians tried to block Sharon from entering the Muslim holy area; extensive rioting ensued and there were clashes with Israeli security forces and a somewhat excessive police response to the riots. Thirteen Arab protesters were killed.

Then, on 29 September 2000, an IDF border police officer was killed by Palestinians. The al-Aqsa Intifada had begun.[42] This revolt involved greater violence by Palestinians against Israel but also a period during which violent state response reflected some of the more kinetic counter-terrorist operational tactics deployed by Israel. Many lives were lost. During 2000–8, Palestinians killed nearly 1,000 Israeli civilians and security force members, and Israeli forces or Israeli civilians killed nearly 5,000 Palestinians.[43] Hamas suicide bombings and rocket attacks formed part of this cycle, with martyrdom attacks often occurring in direct response to Israeli actions. During this Second Intifada, violence came to overshadow non-violence as the Palestinian method of resistance.

These activities by Hamas reduced sympathy for all such groups, while international engagement remained a notable and important aspect of the conflict. In April 2001, the Sharm el-Sheikh Fact-Finding Committee (chaired by US Senator George J. Mitchell, who had been prominent in the Northern Ireland peace process) delivered its Report to President Bush. This argued that the Israeli government and the PA needed to work urgently to resume negotiations, to restore trust and confidence, to cooperate on security matters, and to halt violence. Mitchell and colleagues rightly alluded to what had been achieved in peace work to date, from Madrid to Oslo and beyond.[44] But this well-intentioned, empathetic, balanced Report was unable to lead to a reduction in terroristic violence between increasingly divergent actors.

Israeli counter-terrorist methods undermined the image of that state, not least because of the seeming transgression of human rights so often involved. With numerous terrorist attacks on Israel, the state called up large numbers of reservists and initiated Operation Defensive Shield in March 2002. Soldiers entered city centres and refugee camps in order to quell terrorist attacks, and Israel reasserted control over some areas that had been ceded to Palestinians.

Operation Defensive Shield did not eliminate Palestinian assaults. But— along with a security fence between Israel and the West Bank, construction of which began in the summer of 2002—there did emerge some tactical-operational progress in terms of thwarting terrorist attacks. During late 2002 and 2003, the percentage of foiled rather than successful suicide attacks grew markedly, and 2004 saw a 45 per cent decrease in the number of Israelis killed by terrorism as compared with 2003.[45] It was not that the desire to attack Israel had evaporated. But terrorists' capacity to do so had been reduced.[46]

Indeed, the year-by-year figures for terrorist-generated Israeli fatalities did suggest an increasing effectiveness in counter-terrorist operational tactics

(2002 (426); 2003 (199); 2004 (109); 2005 (50); 2006 (24)).[47] This life-saving tactical-operational work did not mean that terrorist assaults ceased. Defensive measures had made it more difficult for Palestinian terrorists to gain access to Israel for suicide attacks, and so rocket attacks were adopted. In 2005, 286 rockets and mortar shells were launched from Gaza into Israel; in 2006, the number was 1,247; in 2007, 938 were launched, as were 1,270 in 2008.[48] Israel continued to respond, both with tactical-operational defensive measures but also with attacks from sea and air or by artillery fire.

These engagements—terrorist and counter-terrorist alike—occurred against a backdrop of continued efforts at strategic or partial strategic resolution of the conflict. In April 2003, President George W. Bush set out his own view regarding the best route towards Middle Eastern progress. This called for the Palestinian Authority to make an unequivocal statement of recognition of Israel and an equally clear commitment to end terrorism and to work to achieve that complete ending; the PA was also asked to commit to political reform. For its part, Israel was called on to declare a commitment to a two-state solution, to withdraw to pre-Intifada territorial lines, to dismantle recently constructed outposts, to cease settlement construction, and to desist from house demolitions. Both Yasser Arafat and Ariel Sharon made positive noises in response to this initiative, but neither pursued it rigorously.

Part of this related to the global counter-terrorist context. After 9/11 and amid the War on Terror, Israel could more easily lump Hamas together with other terrorists internationally (despite the profound differences that exist between Hamas and al-Qaida).[49] The US was now committed to the eradication of terrorism throughout the world, to seeking allies in this struggle, and to supporting democratic states; Israel's hand was strengthened as a consequence.

The tensions between Israeli tactical and strategic/partial strategic efforts are important to note. It was understandably difficult to sustain Israeli popular support for a peace process with Palestinians while bombings and attacks were still ongoing. For some Israelis, a deal had been done, concessions had been made by Israel, territory had been yielded, recognition had been given to the PLO, Palestinian prisoners had been released—and yet, after all of this, terrorist attacks and tragic deaths still persisted. This was the paradox of peace-process counter-terrorism. To end Palestinian terrorism decisively would require a strategic deal such as Oslo; but to end the daily experience of terrorism in Israel was made more difficult while Hamas and others reacted to Oslo by engaging in terrorism. Stopping the latter terrorism was harder to achieve amid violent responses to the former deal, and because Israel was not now in control of Gaza or the West Bank (and so could not enforce measures against terrorism there or sustain intelligence-gathering activities to the same degree). Moreover, the politics of peace deals meant that Israel was wary of tactical actions that might aggravate relations with Palestinians with whom the deal was being sought.

This reflected a longer-term difficulty in Israeli counter-terrorism. The state did tend to pursue both strategic/partial strategic success and tactical success at the same time, and for understandable reasons. Ultimately, only the former would bring the threat of terrorism to an end, but there was a need to protect life and limb from ongoing daily attacks. The problem was that the two approaches could clash with one another and frequently did.

At times, of course, they could also work in harmony. Although the Oslo peace process ultimately disintegrated without a resolution to the Israeli/Palestinian conflict, the strategic/partial strategic progress of that period should not be ignored. This is true in terms of the eye-catching agreement reached between previously fierce adversaries. But there was also a tactical dimension. The newly established Palestinian Authority did have a containing effect on Hamas violence, and the numbers of Israelis killed by Palestinian terrorism significantly diminished during the 1990s phase of development (fifty-six were killed in 1996, forty-one in 1997, sixteen in 1998, and eight in 1999). Not unrelatedly, the numbers of Palestinians killed by Israel in the Occupied Territories also diminished: whereas 290 Palestinians had been killed by Israeli forces in 1988, 112 were killed in 1994, and 18 in 1997.[50] To this extent, strategic compromise (however partial), and the building of more positive relations between Israel and a significant section of Palestinian politics, had for a time produced one of the key outcomes to be desired by effective counter-terrorism: the reduction of fatalities.

In the longer term, however, the emergent ascendancy of Hamas rather than the PLO within Palestinian resistance efforts prevented such developments from growing stronger. Hamas were much more powerful in Gaza than they were in the West Bank, and they were not the only Palestinian group for Israel to deal with. Islamic Jihad had been formed in Gaza in the early 1980s, began attacks against Israelis during 1985–6, and engaged in numerous lethal assaults against their Jewish enemies. But the ultimate stalling of the peace process reinforced the leading role of Hamas in anti-Israeli resistance. This involved more than terrorist violence. Indeed,

Hamas's political and social strength complemented their aggressive methods in ways that caused great concern to Israeli authorities.

This became even more salient after Israel's departure from Gaza in 2005. During August and September of that year, Israel withdrew its settlers and soldiers from Gaza; by mid-September, all Israeli settlements there had been evacuated and destroyed, and the IDF had left the area. Evacuated territory was then seized by Hamas (and shortly after the withdrawal, there were rocket attacks from Gaza upon Israel). This withdrawal occurred unilaterally, without any deal. Hamas presented it as a success brought about by their own violent struggle. Once Israel withdrew, Hamas duly became the most significant player in Gaza. The Israeli departure had been a controversial move. It affected what Israel could in practice do in terms of counter-terrorism in Gaza, and removing Jewish settlers from their homes there was painful for some Israelis to witness. The withdrawal meant that Hamas would be able to take over Gaza, diminish its rivals there, and impose its rule and cultural preferences. It also facilitated its continued use of the area as a base for terrorist attacks on Israel, via tunnels and rocket attacks.

Gaza therefore became a famous Hamas stronghold, and one in which it could enforce and oversee strict Islamic rules of conduct. High population density, extensive unemployment, and great desperation and humiliation provided foundations for angry resistance to Israel. In this context, Hamas transitioned towards being a resistance movement which was also a power-administering organization.

In the Palestinian Legislative Council elections of 26 January 2006, Hamas won 74 of the 132 seats (Fatah won only 45). In part, this represented a vote against what was perceived as the more corrupt Palestinian Authority. But it also reflected support for the supposed efficacy of Hamas violence, and the fact that Hamas's more intensely anti-Israeli approach resonated with many voters. Hamas then brutally took control of Gaza in June 2007, when there was extreme violence between Hamas and Fatah (the former expelling the latter from the area). Hamas was now in government and simultaneously an active terrorist organization.

Israeli counter-terrorism consequently involved tactical-military methods on frequent occasions. In December 2008, Israel launched a twenty-five-day attack on Gaza in response to Palestinian rocket attacks; at least 1,300 Gazans were killed in this Israeli offensive and over 5,000 were wounded, most of them civilians. Operation Cast Lead, the name given to this 2008–9 endeavour, was presented as embodying both defence and deterrence. In practice,

Israel's comparative strength was such that there were huge losses on the Palestinian side through this large-scale military operation. The attacks began on 27 December 2008 with an Israeli Air Force assault, intended to immobilize the Hamas infrastructure that was used to launch rockets into Israel. Heavy damages were suffered by Hamas, but the attack did not immobilize their capacity to counter-attack. Operation Cast Lead ended on 17 January 2009, after three weeks of conflict. Thirteen Israelis had been killed, and there was considerable international criticism of Israel following the operation, not least because of the profound asymmetry of the fatalities.

Again, therefore, there were tensions between different kinds of counter-terrorist success or failure. Operation Cast Lead did allow for some inherent rewards in hitting back at Hamas for its own assaults on Israel. It did do damage to Hamas, undermining some of its capacity. Hamas targets were indeed hit, and rocket attacks on Israel during coming years did drop: in 2008, 1,270 rockets had been fired into Israel; in 2009 (following Operation Cast Lead) the figure was only 158; and in 2010 the figure fell to 103.[51] In terms of publicity, however, the operation was a negative development for Israel, reinforcing for many observers the state's tendency towards disproportionately harsh responses. The pain inflicted on Gazan civilians understandably formed part of this. It is true that Hamas did use civilian areas to store weapons and also used civilian homes as bases from which to attack Israel.[52] But, while this helps explain why Israel acted as it did, few observers considered it to justify the intensity of the force that Israel deployed. Moreover, public support for Hamas violence seems to be highest when there appears little hope of conflict resolution or of the achievement of peace.[53] In this sense, an intensified experience of Israeli attacks is unlikely to undermine or erode the support base of Israel's main terrorist opponent.

Hamas was not, of course, Israel's only terrorist adversary. The Lebanese Hezbollah ('Party of God') had emerged in 1982. Becoming more strongly organized and coherent by the mid-1980s, the organization had major sponsorship from Iran and interpreted its violence as Islamic resistance against Israel.[54] There is no doubt that Hezbollah has carried out acts that would fit most definitions of terrorism[55] but, as also with Hamas, Israeli counter-terrorism was complicated by the fact that the organization also acted in areas outside terrorism. For Hezbollah became a substantial military but also political opponent.

In May 2000, after long occupation, Israel had withdrawn from southern Lebanon following resistance from Hezbollah, which enjoyed considerable support from Shia Muslims. Hezbollah's role in politics (including electoral politics), and its significance in providing social services for its population,[56] have reinforced its support with many in Lebanon. As so often, therefore, Israel's responses to terrorist acts were made more difficult because of quasi-democratic aspects to its opponents' work.

Conflict frequently became inflamed. On 12 July 2006, Hezbollah attacked Israeli border towns with rockets, and took two Israeli soldiers prisoner on the border between Lebanon and Israel (with the aim of exchanging them for Lebanese prisoners then held by the Israelis). During July and August, there was a fierce conflict between the IDF and Hezbollah, with the latter being heavily hit. Neither side cleanly won this conflict. In terms of deaths, infrastructural damage, and economic calamity, the war did huge damage to Lebanon;[57] in terms of international opinion and publicity, both sides faced criticism for the targeting of civilians. For Israel, the key difficulty was the ongoing endurance of this major adversary, so anti-Hezbollah counter-terrorism had to persist. On 12 February 2008, for example, Israel killed Hezbollah leader Imad Mughniyeh in Syria. But this kind of attack did not end Hezbollah's existence as a major opponent for Israel.[58]

Gaza and Hamas likewise represented ongoing challenges, to which Israel episodically responded with very aggressive tactics. In February 2008, following rocket fire into Israel from Gaza, the Israelis sent another ground incursion into the area. Operation Hot Winter saw Israeli infantry enter the northern Gaza Strip for a forty-eight-hour operation aimed at damaging terrorist infrastructure. In 2012, another escalation of anti-Israeli terrorism prompted a counter-terrorism response (Operation Pillar of Defence, launched in November). The tactical aims here were to strengthen Israel's deterrence against terrorism emanating from Gaza: specifically, to damage Hamas's rocket-launching capacity; more generally, to undermine Hamas and other terrorist organizations with the goal of providing greater security for Israel. The eight-day operation mostly involved air strikes against Hamas targets, saw Hamas and Islamic Jihad continue to fire rockets at Israel, and resulted in fatalities (six Israelis, and nearly two hundred Palestinians). There does seem to have been some tactical-operational effect. In 2013, seventy-four rockets or mortar shells were launched into Israel from Gaza; in 2012 (excluding the attacks during Operation Pillar of Defence itself), 280 rockets had been launched.[59]

The rhythm of such terrorist/counter-terrorist engagements owed much to fluctuating levels of attack on Israel. But the broader context of Palestinian desperation is crucial too, as is the sense possessed by many Palestinians that there must be resistance to the Occupation—meaning the contemporary occupation by Israel of Gaza (from 2005 onwards, from Gaza's outer perimeter), of the West Bank, and of East Jerusalem. Palestinians interpret this as the Israeli occupation of what is rightfully Palestinian land; the consequent effects of the occupation on the daily lives of Palestinians are therefore seen as Israel's responsibility. For their part, Israeli politicians have sought that any peace deal should come with practical, immediate benefits in terms of a cessation of Palestinian terrorism. As Prime Minister Benjamin Netanyahu put it in December 2013, 'The state of Israel, I believe, has a strategic interest in diplomatic negotiations aimed at reaching an agreement that will end the conflict.... A diplomatic agreement will be signed only if these vital interests are secured, first and foremost our security and their demilitarization.'[60] So the daily experience of both Palestinians and Israelis has somewhat limited the scope for long-term, patient commitment to conflict-resolving politics.

In 2014, Israel's Operation Protective Edge killed more than 2,000 Palestinians (most of them civilians) and wounded more than 10,000.[61] The repeated Israeli attacks on Gaza were justified by Israel in terms of defence and deterrence, but the scale of loss of life remains shocking and emblematic. Launched on 8 July 2014, Operation Protective Edge followed an escalation of rocket attacks into Israel from Hamas and other Palestinian groups in Gaza. The Israeli goal was to limit and deter Hamas and others from engaging in such terrorism. There were fifty days of fighting, and numerous tunnels were destroyed by Israel, along with many other targets in Gaza. In terms of Israeli defensive measures, considerable technical sophistication has often been involved. 'Iron Dome', for example, is an Israeli system aimed at intercepting rockets over an extensive area, and has proved effective against Hamas attacks.[62]

The broad pattern outlined above—Israeli strikes against Gaza-based terrorists; Palestinian rocket attacks on Israel; the killing of militants by the Israelis, but also collateral damage in the form of civilian Palestinian deaths and lower-level casualties in Israel—has been a depressingly predictable cycle.[63] Moreover, in the face of increasing defensive measures by Israel, terrorist tactics could prove fluid. Lone-actor Palestinian assaults in Israel also became a more common phenomenon (these include stabbings, as well as vehicle-based ramming attacks). Encouraged by organizations like Hamas, 2015 saw attacks of this kind in Israel, the attackers not formally a part of terrorist organizations. Despite the professionalism and commitment of Israeli counter-terrorists over many years, the terrorist threat persisted.

7
Tactical Successes, Strategic Failures?

I

Much denunciation of Israeli counter-terrorism avoids recognition of the considerable variations in approach that have been adopted by different regimes (and different Prime Ministers) during Israel's complex past. In understanding Israeli counter-terrorism, one needs to recognize change as well as continuity over time and also to adopt the kind of framework deployed in this book for analysing the different ways in which counter-terrorism might variously be judged to have worked or not. Israeli counter-terrorism has been heterogeneous rather than homogeneous, as is reflected in the suggestion that the late 1960s aim of eliminating terrorism was followed by the early and mid-1970s goal of deterring terrorists from attacks, and then by the late 1970s and early 1980s intention of minimizing the damage done by a terrorism which could not be eradicated.[1]

Much of what has happened in Israeli Counter-Terrorism has embodied tactical responses to a terrorism from which Israeli governments and people have simply wanted to protect people on a daily basis. In itself, the tactical goal of making the terrorist flame burn less strongly is a major and understandable one, and so much Israeli attention has focused on this tactical level of work and on the attempted limiting of terrorist attacks. Accordingly, Shin Bet (or Shabak, the Israeli internal security agency)[2] has frequently approached counter-terrorist decisions primarily on the basis of their likely operational and immediate effectiveness. Between intelligence-based prevention, the toughening of defensive measures, the capture and detention of terrorist activists, and targeted assassinations, Israel has at times had considerable tactical-operational success in diminishing the number of

Palestinian terrorist attacks against it.[3] And the degree of professionalism, acumen, and skill evident here has at times been remarkable. One of Shin Bet's most significant assets (their agent Mosab Hassan Yousef, the son of a leading Hamas figure) noted the extent to which Israeli counter-terrorists were able to gather vast information about, and a genuinely intimate understanding of, their Palestinian enemies.[4] Great operational skill and professionalism have repeatedly been shown by Israeli counter-terrorists.[5]

There are, however, complexities, and these are starkly reflected in the case of targeted assassinations. Such killing of terrorist opponents has long been one of the tactics used as part of the Israeli state's counter-terrorist repertoire. Two kinds of target have particularly been involved: people thought to be about to launch an attack; or senior terrorist players, those who plan and direct the strategy. For the counter-terrorist, assassinations combine the attractions of prevention, deterrence, organizational attrition, and punishment. But there is the danger of generating a violent reaction too, and the latter might take two forms. There can be immediate cycles of revenge; but there is also the possibility that such targeted killings (which sometimes cause collateral damage) might confirm your enemy's perception of you as a violent oppressor and might therefore help to sustain longer-term resistance and violence against Israel.

Sometimes, targeted killings embody an attempt to do something decisive in the wake of a particular atrocity. In October 1994, a terrorist attack in Tel Aviv killed twenty-one people; after this, there was a strong impulse to target terrorist leaders, and a senior Islamic Jihad activist in the Gaza Strip was indeed killed.

The number of such targeted attacks has been considerable. Israel attempted 159 targeted killings between September 2000 and April 2004, with a success rate of 85 per cent. Many of these targets were Hamas people, from different levels of the organization,[6] and although Hamas was still able to operate militarily during and after these years, the tactical-operational successes of Israel were notable over a long period. Hamas leadership losses included: Salah Shehadeh (head of Hamas's military wing), killed by Israel on 22 July 2002; Sheikh Ahmad Yassin (Hamas's founder and spiritual leader), killed by a helicopter strike on 22 March 2004; and Abdel Aziz al-Rantisi (Yassin's successor), killed on 17 April 2004. Such actions could also, however, produce counter-productive results. Although prior terrorist activity was the basis for these targeted assassinations, lethal responses by Hamas could be triggered by them too. On 31 August 2004, Hamas carried out a double suicide bombing in retaliation for the Yassin and Rantisi killings, killing sixteen civilians. This kind of cycle had longer roots within the mutually shaping relationship between Israeli counter-terrorism and Palestinian terrorism. On 5 January 1996, Hamas bomb-maker Yahya Ayyash ('the Engineer'—one of Hamas's military leaders and someone involved in the organization's ongoing suicide-bombing campaign), was killed in Gaza by Shin Bet; the latter had recruited a family member of one of Ayyash's Hamas colleagues as an agent, thereby making the assassination possible. Numerous attacks by Hamas during February and March 1996 ensued in direct retaliation for Ayyash's killing, leading to the deaths of more than forty Israelis.

As this grisly sequence might suggest, there is disagreement regarding the efficacy of Israeli targeted killings. Some have claimed that, together with other measures, targeted killings can help reduce the number of terrorist attacks; and the creation of strong defences, and success in arresting other terrorists, are necessary to complement more lethal operations.[7] Here, it might be noted that the Israeli arrest of Sheikh Yassin and around 250 activists in May 1989 had itself represented a serious tactical blow to Hamas. More broadly, indeed, it seems that preventative arrests probably possess greater efficacy in limiting terrorism than do targeted killings themselves.[8]

Brutal though they are, targeted killings are probably most effective when they are selectively used. It is crucial to target those who will be seen by many as most deserving, it is also vital to avoid collateral (civilian) damage, and it is important to target those whose death will in practice limit the number of subsequent terrorist attacks. Achieving this is far from easy, even for as adept a state as Israel. Some claim that Israel's leadership decapitation of Hamas has had tactical efficacy in constraining Hamas's capabilities and weakening the organization,[9] though even those who consider targeted killings successful can acknowledge that such tactical effectiveness still leaves the strategic problem that terrorism continues to persist.[10] But the effects on other aspects of tactical counter-terrorism have been less positive, and it is necessary to consider the full range of consequences. There has clearly been very negative publicity generated by the extent of civilian harm done by these Israeli attacks, with considerable undermining of Israel's global reputation. Moreover, in terms of sustaining control over a population, the terrible pain inflicted on Palestinians in Gaza has, like Israeli military engagements in Lebanon, offered motivations for popular sympathy towards Israel's violent enemies.[11]

Something similar can be seen in relation to home demolitions, deployed on occasions by Israel as a collective punishment for Palestinian terrorist attacks. Some scholars suggest that demolitions do not possess a deterrent effect.[12] There is also some evidence of short-term reductions in Palestinian attacks being achieved but, again, there is also evidence that the policy has prompted an even further intensification of anti-Israeli anger, thereby sustaining the terrorism that the policy was intended to diminish. As so often in counter-terrorism, the more selective the targeting, the more likely it is that it will prove effective.[13] Yet even if house demolitions do possess some operational efficacy, the publicity effect on international opinion has been harmful. Together with curfews, checkpoints, separation barriers, deportations, and restrictions on movement, there has been a widespread image of repressive day-to-day Israeli occupation. Effective counter-terrorism partly depends on sustaining credibility with relevant audiences. In terms of international opinion, this has proved a repeatedly difficult challenge for Israel in the wake of its more aggressive counter-terrorism engagements, however operationally successful they might have been.[14]

International opinion, of course, can have tactical as well as strategic implications. Benjamin Netanyahu, somebody whose career has involved strong commitment to countering terrorism, has stressed the importance of sharing intelligence: 'One of the central problems in the fight against international terrorism has traditionally been the hesitation of the security services of one nation to share information with foreign services.... Only through close coordination between law enforcement officials and the intelligence services of all free countries can a serious effort against international terrorism be successful.'[15]

In terms of tactical-publicity work, the problem of international perceptions has certainly been longstanding. In 1986, two authors could suggest that international terrorism had 'succeeded for the Palestinians. Public opinion in the West has found it increasingly difficult to ignore the threatening poses of Palestinians carried in newspapers and on television, and traditional sympathy for Israel has been to some extent modified by calls for a peaceful solution of the Middle East conflict that recognizes the PLO as the legitimate representatives of the Palestinian people.'[16] Sympathy for Israel has also been diminished by some of its muscular tactics, as outlined above.

There is evidence here, therefore, both of different tactical outcomes existing in tension with each other (tactical-operational versus tactical-publicity outcomes, for example), and also of tactical successes possibly making Israel's strategic position much more challenging. As in many other counter-terrorism contexts, it is possible for tactically effective work to intensify and sustain the energy behind the terrorist threat that such work was intended to extirpate. Protecting citizens from terrorism surely represents a legitimate and high-priority goal for the Israeli state. But securing those daily tactical successes can repeatedly involve counter-productive tensions. Sometimes (as when operational success allows for interim concessions such as prisoner release for Israelis),[17] different tactical engagements can work harmoniously with one another. On other occasions they can obstruct each other's effectiveness, and tactical successes can sometimes make strategic success more elusive.

Part of this involves tensions between the different political pressures that are involved. On the one hand, there are liberal-democratic values and international influences that push Israel towards restraint; on the other, there is a domestic demand for action, as well as the fact that some of the more successful operational methods lead away from restrained counter-terrorism.[18] One challenge here is that the war of narratives and publicity can exist in conflict with the physical struggle against terrorist action. Propaganda, popular sentiment, and the public context are crucial elements in counter-terrorist success. Effective Israeli counter-terrorism against Palestinian organizations needs to involve far more than merely tactical-operational work. The struggle for legitimacy within and beyond the region is also vital, as are inter-related social, economic, and media-focused engagements.[19]

In terms of the tactical work of undermining opponents, how successful has Israel been? Some Israeli counter-terrorism, including offensive action against terrorist organizations and their supporters, has been aimed at damaging the morale of the state's opponents, through making people experience high costs for resistance, and through undermining confidence in leadership capacity or in the chances of terrorist success.[20] It is evident that a sense of desperation and humiliation do prevail among some Palestinians, though it is also true that support for anti-Israeli militants has persisted.[21]

Likewise, the tactical concern to control population has only been ambiguously addressed. Israeli policy has involved sustained support for Jewish settlements. By 2006 in the West Bank/Judea and Samaria, for example, the number of Jewish settlers had grown to over 250,000; by 2016 the number of Jewish settlers in the West Bank and East Jerusalem exceeded 600,000.[22]

Has all this counter-terrorism produced an Israeli organizational strengthening, another layer of possible tactical success? There is no doubt about the investment in, and sustained resilience of, the counter-terrorist wings of the Israeli state. Defensive structures and policies have fortified both state and society. Tactical defensive measures have included the barrier between the West Bank and Israel, which has helped make suicide attacks much more difficult. Security checkpoints have made it much harder for Hamas and other terrorist organizations to attack Israel, as have patrols, roadblocks, and the reinforced defence of buildings. Indeed, it has been strongly argued that Israel's most significant successes in counter-terrorism have involved precisely the development of defensive measures as a means of protecting citizens.[23]

These security measures are expensive. In the late 1990s, the cost of security for Israel's national airline El Al—a cost mostly paid for by the government—was around $80 million per annum.[24] Just as Israeli intelligence has led to the repeated prevention of planned attacks and to other means of disrupting terrorist adversaries, so too Israel's defensive measures (such as aeroplane security or border security barriers) have made terrorist attacks harder to achieve.[25] And effective counter-terrorism does necessitate that the state avoid its people's demoralization in the face of terrorist attack. Moreover, Israel has indeed persisted as a highly successful and resilient society in terms of daily life, economy, and cultural achievements.[26]

There has remained an explicably deep sense of Jewish vulnerability and insecurity, owing to the ongoing violent assaults by terrorists, the hostility of regional neighbour states, and the wider context of traumatic inheritance encapsulated by novelist Philip Roth as 'the mayhem of Jewish life in the twentieth century'.[27] Terrorism itself has certainly done some damage to the Israeli economy, with negative effects on businesses such as restaurants and cafes at times of intense threat. Still, resilience has been broadly and strongly sustained. And Israel has demonstrated a profound counter-terrorist commitment and a capacity to ensure its own survival—crucial signals to send in effective counter-terrorism.

US support for Israel has been significant here. The Israeli/Palestinian conflict has simultaneously involved rivalry between nationalisms in the local context, and also both wider-regional and even global competition. Many Palestinians have had an understandable anxiety regarding the US's strongly pro-Israeli approach, and very strong US support has indeed represented a vital component of Israeli counter-terrorism. Israel has been that rare thing, a Middle Eastern democracy; it was also a Cold War ally of the

US against the Soviet Union; and there exists a very strong pro-Israel lobby within American politics. The US provides vast financial and military backing to Israel, and as such is seen by many Palestinians as being far from a neutral arbiter.[28] For many people supportive of Islamic terrorism (and not just that of Hamas), US support for Israel legitimizes anti-American violence; so in reality US foreign policy here has been crucial to anti-US terrorism.[29]

II

Tactical activity has involved the securing of some inherent rewards through Israeli counter-terrorism. Yitzhak Shamir (1915–2012; Israeli Prime Minister during 1983–4 and 1986–92) commented on the punishment of terrorists, in order that they should 'get what they deserve'; it was important, Shamir said, to 'show these bastards who kill us mercilessly that there is someone who knows how to strike back. We have to prove to all these murderers [that] we will take revenge on them in any way we see fit'; 'We can prove to the terrorist organizations and their leaders . . . and to all those who are involved in Palestinian terrorism, that acts of terrorism will harm them more than they harm us, for example by way of demolitions of the murderers' homes and the expulsion of all those who collaborated with them.'[30] The tactical intention to deter was here very clearly interwoven with the emotional dynamics of hitting back in vengeful anger.

In light of this we could say that there has been some emotional satisfaction achieved through, for example, house demolitions as a way of striking back in revenge for terrorism. There also has been, for some people in state forces and in the wider population, an improvement of morale after targeted killings: these can seem to satisfy a demand for punishment and revenge, just as the threat of such targeted attacks forces terrorist adversaries to devote energy and time to self-protection.[31] It is also true that Israeli politicians can gain inherent rewards through the authorization of targeted killings: there is often a popular desire for hitting back at those who have attacked Israel and its people and a sense that—after appalling terrorist atrocities—something must be done.[32] More broadly, electoral support for Israeli right-wing politicians has tended to grow after periods of intense terrorist violence.[33] And politicians' careers could benefit greatly from counter-terrorist successes.[34]

The Israeli state is composed of human actors whose emotions are understandably engaged by what counter-terrorism involves.[35] During the 2008–9 Operation Cast Lead against Gaza, some IDF soldiers seem to have gained such benefits, if we look at their testimonies (which include the mention of 'joy' or 'delight').[36] Through counter-terrorist operations, military personnel could experience pride and excitement, and they could gain considerable renown, but they could also suffer danger, loss, serious injury, and death.[37] Beyond the state itself, those who became informers for the Israelis could see themselves as performing the righteous task of saving people's lives and could relish the work.[38] And for others too there have been benefits. In contrast to some sectors, defence-related companies in Israel have benefited economically from the effects of terrorism, with opportunities growing considerably for business in the security industry.[39]

Relations between the tactical and inherently rewarding levels of counter-terrorist success have not been unproblematic. Some tactics have worked against the efficacy of others, and some emotional rewards might be derived from engagements which have a questionable overall benefit for Israel. Indeed, while limiting attacks (tactical-operational success) is clearly important, it is strategic or partial strategic success that will lastingly and ultimately limit violence against Israel, and here the complexities of counter-terrorist experience become very clear. At its most basic, the point can be expressed by recognizing that much terrorism is sustained through a desire for revenge, and that robust Israeli counter-terrorism (targeted assassinations, air strikes, house demolitions, and the like) provides just the kind of pretext for Palestinian violence leading to tit-for-tat cycles of retaliation.[40] One former Prime Minister of Israel declared his and his governmental colleagues' responsibility as being 'to care for the future of the Jewish people, for the safety of the Jewish state, for the security and the lives and the liberty of our children'.[41] But tactical attempts by successive Israeli governments to protect security and daily lives have frequently proved counter-productive in relation to the longer-term strategic objective of protecting the state and the future of its people.

III

Ultimate Israeli counter-terrorist success would involve the eradication (or the partial yet substantial eradication) of the basis for terrorism itself.

Different timelines are involved here, and that is part of the problem. Reinforced by short-term electoral pressures, governments feel the need to do something practical (and tactical) in order to limit daily terrorist violence, and to be seen to do something strong in defence of their people. By contrast, peace-process strategy requires long-term, patient investment in processes that can be accompanied by some ongoing violent terrorisms. Understandably, Israelis (politicians and citizens alike) have wanted to see immediate security, while pursuing longer-term peace. Israel's Prime Minister Benjamin Netanyahu, speaking in 2013, put it emphatically: 'We insist upon security. Without security, there can be no agreement. Peace that cannot be defended is not peace. I will not accept trickles of missiles, rockets, or terrorists.'[42]

This book's framework emphasizes the many different things simultaneously at work in counter-terrorism, and the numerous ways in which one can define success. The challenge (faced by Israel more than most states) is a pervasive terrorism which demands a strategic resolution and which simultaneously makes that seemingly impossible to realize.[43] Ami Ayalon (Commander of the Israeli Navy between 1992 and 1995, and head of Shin Bet between 1996 and 2000) candidly identified the danger: 'War against terrorism is part of a vicious cycle. The fight itself creates...even more frustration and despair, more terrorism and increased violence.'[44] Part of the issue here has been the frequently very militaristic nature of Israeli responses to terrorism,[45] and the fact that Israel has often proved stronger at counter-terrorist tactics than it has been at longer-term strategy.[46]

Tactical successes should be recognized for both their impressive qualities and their limitations. Amid the latter, Israel's targeted killing of leaders, alongside other heavily military actions, have provided grievances prompting many Palestinians to support the terrorism which these actions were supposed to prevent. The narrative war has consequently often gone badly for Israel, with much international propaganda, argument, and opinion heavily critical of the state's response to the terrorist challenge. Sometimes, short-term advantage has been pursued without sufficient consideration of long-term implications, not least with regard to the possibility of a decisive deal to bring most terrorism to an end; and Israel's successes in dealing with Jewish terrorism have not been sufficient either to end this blood-stained phenomenon or to convince Palestinians that Israel is committedly hostile to such pro-state terrorism.[47] Ideally, states require alignment between strategic/partial strategic success, tactical success, and inherent rewards; in practice, in relation to Israel, this has proved markedly elusive.

Israel's considerable success in countering particular tactical threats has therefore sometimes worsened the cyclical endurance of the conflict from which future terrorist threats will emerge. In terms of intelligence-gathering, attack prevention, broader defensive measures, and the operational damage done to terrorist enemies, Israel has achieved much. But the motivation for ongoing Palestinian terrorism has sometimes been strengthened by Israeli tactical actions which have (in the shorter-term) constrained particular dangers.[48] In short, we have tactical success and strategic failure.

There are some who hold that the conflict with the Palestinians will never end, and that the right thing to do is therefore not to focus on strategic victory but rather to concentrate on tactical successes (deterrence, attack prevention, containment, target protection). In contrast to this, it might be pointed out that many people on both sides of the divide have shown a preference for a peace deal, and also that groups like Hamas do respond to popular views among their base.[49] The normality of most of those who engage in and support terrorism[50] reinforces here the possibility that at least partial strategic success in eroding the basis for terrorism might yet be attainable. During the early 1990s it seemed to many that the Israeli/ Palestinian peace process would survive, but that such a deal would prove unattainable in Northern Ireland. These complex processes are contingent rather than inevitable, and the near-fatalistic assumption that violence-sustaining tactical operations represent the best counter-terrorism available to Israel is therefore perhaps unduly pessimistic.

Indeed, unless Israel reaches strategic or partial strategic counter-terrorist success, a far worse situation for the state might yet develop. Complementing the difficulties that exist with Palestinians in Gaza and the West Bank, there are also deepening tensions within Israel itself, where polarization between the Jewish majority and the Palestinian/Arab minority has greatly intensified.[51]

One crucial aspect of all this is that terrorism and counter-terrorism are not the truly main issues here, eye-catching though they have been. Rather, these are terrible symptoms of the main phenomena involved: rival, religiously influenced nationalisms; questions of state legitimacy and of inter- and intra-state relationships and power; borders, territory, security, and identity. If Israel is to avoid ongoing fire-fighting and tactical struggle, then strategic or partial strategic successes need to be aligned with tactical and inherently rewarding work.

Israeli/Palestinian peace-building has proved extraordinarily difficult. But there are grounds for thinking that major terrorist organizations such as Hamas might perhaps find strategic value in shifting their work away from the violent and towards the more constructive.[52] It still remains possible that a two-state resolution could be reached in Israel/Palestine[53] and, while this would be unlikely to eradicate all political violence, it might prove the long-term basis for decisively lowering levels of terrorism and could be considered a partial strategic success if it did so.

IV

We have so far mainly focused on Israeli counter-terrorism in relation to Palestinian violence. But pro-state, Jewish terrorism has been a repeated theme in Israeli politics and embodies its own serious dangers. As noted, terrorism had formed part of the campaign to establish the state of Israel in the first place,[54] and there were also Jewish terrorist groups founded after the state came into existence. Where Israel has taken a pro-settler stance, it has sometimes been with a view to restricting Jewish terrorism and, as in Northern Ireland, the state's efforts to deal with anti-state terrorism have partly been constrained by issues relating to the effect of state action upon levels of pro-state violence. Dealing with Palestinian and Jewish terrorisms at the same time has been an extraordinarily difficult task. Moreover, too gentle or too harsh a reaction to Palestinian terrorism each carries with it risks for the Israeli government.[55] Israel has tried negotiation, targeted assassinations, the invasion of neighbouring states, protective defence mechanisms, curfews, imprisonment, retaliatory violence, and much else. The state of Israel still exists, and strongly so, and there have been repeated tactical successes. But the elusiveness of strategic or partial strategic success remains a profound problem for this troubled democracy.

Some of Israel's enemies, whether neighbouring states or non-state terrorist adversaries, have sought that Israel be destroyed. Whether terrorist groups have ever approached the level of strength to enable them to deliver this goal is doubtful.[56] But the appalling twentieth-century Jewish experience of attempted annihilation together with this ongoing shadow of malevolent intention and repeated attacks have between them made the defence of the state against violent opponents an understandably high priority. The PIRA and ETA (Euskadi Ta Askatasuna—Basque separatist group) respectively wanted the UK and Spanish states to redraw their boundaries by ceding sovereignty over some territory; they did not seek that those states should be replaced across their entirety, that the states themselves should be destroyed. Reinforcing this point is the long-term and deeply painful Israeli experience of personal loss in the face of terrorism and the undoubted tenacity of Israel's committed enemies. As one Hamas leader put it, the issue was 'getting rid of the Zionist presence on this land'.[57] Central, therefore, to effective Israeli counter-terrorism is the demonstration of long-term commitment.

Some observers have assessed Israeli counter-terrorism in largely positive terms. Professor Nadav Morag, for example, examined in detail important factors during the 2000–4 period (levels of civilian casualties among Israelis and Palestinians, Israel's economic performance and social cohesion, levels of domestic and international support for Israeli governments and for Palestinian leaderships), and concluded that, 'The picture that emerges here is one of fairly unambiguous success in terms of every parameter save decreasing the number of Palestinian civilian deaths and the attempted erosion of the late Yasser Arafat's domestic political support'.[58] Morag's article does identify crucial Israeli successes at the tactical level but the period studied is short and so doesn't include analysis of the important question of how far prior Israeli policies had helped to produce an avoidably worse (or better) situation than that which obtained in September 2000, or an understanding of the effects of Israeli actions during that very brief period on longer-term politics and violence.

Professor Joshua Freilich's evaluation of Israeli counter-terrorist efficacy has suggested that there has been a lack of coherent strategy on the part of Israel and that the country's international standing has been damaged by some counter-terrorist actions. But this analysis also concluded that Israel's economy has survived, despite terrorism, and that relative security has been achieved, albeit through ad hoc rather than more sustainedly coherent thinking.[59] The picture painted is one of considerable tactical success (minimizing Israeli casualties, for example), complemented by a failure to make the kind of strategic progress that would more lastingly erode the threat posed by terrorism. On the basis of long-term historical analysis, informed by our fourfold framework for interpretation, this seems a plausible conclusion.

Conclusion

Today, we are a nation awakened to the evil of terrorism, and determined
to destroy it.

President George W. Bush (11 October 2001)[1]

I say that the events that happened on Tuesday September 11 in New York
and Washington are truly great events by any measure, and their repercussions are not yet over. And if the fall of the twin towers was a huge event,
then consider the events that followed it ....

Osama bin Laden (20 October 2001)[2]

In Memory of
Edgar Samuel David Graham
Assembly Member for Belfast South 1982–83
Shot by Terrorists on 7 December 1983
'Keep Alive the Light of Justice'

Edgar Graham Memorial, Belfast[3]

She was, I suppose, the 'apple of my eye'. . . . It is hard to believe, after so
many days and weeks, and now years, of shock, suffering, and loss, that
Marie has gone. Somehow, her presence still permeates the house. There
are some days when you still expect her to fling open the back door, to
burst into the room and exclaim, 'I'm back again! I'm here! What's for tea?'
But she isn't here, and my wife Joan and I have tea on our own.

Gordon Wilson, on his daughter Marie, who was killed
by the Provisional IRA's 1987 Enniskillen bomb[4]

The only thing that matters is that we can exist here on the land of our
forefathers. And unless we show the Arabs that there is a high price to pay
for murdering Jews, we won't survive.

Israeli Prime Minister David Ben-Gurion, 1953[5]

I

The central aim of this book has been to generate constructive, dispassionate, historically informed debate about a subject which is as emotionally charged as it is globally significant. There can be little doubt about the ongoing challenge posed by terrorism nor about the degree to which responses to terrorist violence continue to determine so much political, economic, and cultural experience globally. Counter-terrorism is therefore an unquestionably significant phenomenon. Despite this, there has been less decisive research and debate regarding its efficacy than is needed. Comfortingly simple answers, or overly mechanical reliance on metrics-based assessments, are unlikely to prove persuasive or satisfactory. What is required, is *first* a systematic, layered, nuanced framework within which to assess in a coherent and consistent way what effective counter-terrorism would involve. Strategic, partial strategic, tactical, and inherently rewarding layers of efficacy can sometimes work harmoniously with one another but can on occasions be in conflict. The lens provided by this framework allows us to read and assess counter-terrorism more effectively than would be possible without it. In particular, this framework allows us simultaneously to recognize the positive, negative, and ambiguous achievements of counter-terrorism, and the important relations between these.

Second, as we interrogate actual experience against this framework, we need to engage with case studies in historical context. This book has argued that history is crucial to political understanding, and that interwoven aspects of an historical approach offer distinctive illumination: an attention to long pasts and, by implication, to long futures; respect for the complex particularity of each individual context and setting; engagement with a large range of mutually interrogatory sources, including first-hand materials; appropriate scepticism about overly neat theoretical models of explanation; and a preference for contingency over inevitability when explaining human activity.

The emerging picture has been complex. *Strategic victory* might involve a state effectively removing the threat against it of more than trivial terrorism, and thereby getting rid of (almost all) terrorist violence. This might involve political resolution of the conflict from which terrorist violence has grown. As in so much human activity, strategic victory has seemed elusive in the case studies scrutinized in this book. In the post-9/11 War on Terror, presidential declarations about destroying, eradicating, or bringing an end to terrorism were unhelpfully unrealistic and were predictably unachieved. Historical understanding of pre-9/11 experience (in the US and elsewhere) would have cautioned against such hubris. Strategic success was not achieved in Afghanistan, in Iraq, or more broadly in the War on Terror. Northern Ireland is the case closest to matching this strategic level of counter-terrorist achievement, but partial strategic success probably better fits what occurred there. In Israel/Palestine there have clearly been significant limits to what might be strategically achieved, given the very lengthy, entangled complexity of the enmities involved in that conflict. It is much easier in this case to criticize the participants than it is to steer an effective journey towards the ending of the struggle. Strategic success has proved elusive, as has a deal to end the Israel/Palestine conflict, despite repeated efforts. Counter-terrorist failures can be as illuminating as successes, and this strategic failure has been as striking in Israel's case as have that state's many impressive tactical achievements.

Partial strategic victory, in which a state significantly reduces terrorist capacity and lethality, and substantially manages to protect its people, has been a more common reality. In the War on Terror, there were occasions when the idea of preserving normal life was stressed, and this has been substantially achieved in key Western states. Partial strategic success can also be seen in the realizing of the aim of not destroying but rather damaging and degrading particular terrorist groups. This has been seen with al-Qaida, with the early phase of work against the Taliban, and with ISIS. The more ambitious secondary goal in Afghanistan of nation-building was ultimately a failure, and partly here the problem was a lack of long-termism on the part of the US and its allies. Following the Soviet departure from the country in 1989, the American eye was taken off the Afghan ball, resulting in a later lack of detailed intimacy of cultural and historical understanding.

In Northern Ireland, the combination of counter-terrorist and wider political work substantially ended the terrorist violence of the Troubles. Some terrorist groups and threats remain, and the peace-process deal has experienced repeatedly bumpy periods of acrimony. But lethal violence has largely disappeared, and it seems fair to judge that the UK eventually secured partial strategic success in Northern Ireland. Terrorist capacity was significantly contained, and a partnership-based political deal has largely ended terrorist campaigns. Supporters and opponents of the UK state in Northern Ireland had seen the state commit to long-term resilience, and to the establishment of a largely peaceful life for most people.

In Israel/Palestine, the success of getting the PLO to recognize the Israeli state's right to exist represented a significant achievement. So too did the maintenance of so much normal life despite committedly violent enemies. Yet again, Israel has clearly prevented its terrorist opponents from securing victory—Hamas's goal of destroying Israel, for example, has never seemed even close to being realized. In all of these ways, some partial strategic success has been realized. But the determined resistance and sustained motivation of terrorist groups like Hamas and the PIJ have not been addressed in such a way as to limit terrorism to the degree that was realized in Northern Ireland (or in other perhaps comparably challenging settings such as the Basque Country).[6]

At the *tactical level* (of operations, interim concessions, publicity, the undermining of opponents, population control, or organizational strengthening) there have been many counter-terrorist successes globally in terms of the thwarting of particular operations,[7] or the developing of particular means of protecting people from attack. The strengthening of cockpit doors in aeroplanes, for example, seems to have been an effective means of making terrorist operations much less feasible.[8] In each of the case studies examined in this book, much has been achieved tactically, though some challenges have also persisted. In the War on Terror, there were significant counter-terrorist coordination problems, within and between states; again, earlier US experience had suggested the importance of this problem. There were repeated tactical successes (targeted drone strikes; early-phase invasions in Afghanistan and Iraq; very significant successes against ISIS). But publicity for some counter-terrorist work could be problematic. The return of the Taliban to power in Afghanistan was one important example; prisoner mistreatment represented another; the loss of credibility over dubious claims to justify the Iraq War was a third. Terrorism actually grew as a result of the Iraq War (within and beyond the country itself), and much of the chaos in Iraq emerged because of a lack of practical planning for the post-invasion period. Where tactical successes in the War on Terror were secured, technological sophistication, considerable skill, high-grade intelligence, and successful coordination facilitated the capturing or killing of many opponents and the thwarting of many terrorist operations.

In the much smaller context of Northern Ireland, again intelligence and coordination were the basis for significant tactical successes. What worked best often had deep roots (the effective use of informers and intelligence, for example); what worked badly again had historical pre-echoes (such as the counter-productiveness of military heavy-handedness). Overall, in Ulster tactical successes increasingly prevented many terrorist operations; they saw the infiltrating and undermining of major terrorist groups; and they led to the imprisoning of many militant adversaries. But the tactical failures were also significant: state collusion with loyalist terrorists damaged nationalist confidence in, and also the credibility of, the state (as well as causing terrible human suffering in some cases); internment and other blunders generated lastingly damaging publicity for the state. But the positive aspects of Northern Ireland counter-terrorist tactics were part of what helped to limit the effectiveness of terrorist groups. This tactical process did therefore play a part in encouraging such organizations towards more political, less violent, methods, because these came to be seen as more effective and fruitful.

Israeli counter-terrorism has been repeatedly very successful and skilful, whether in terms of extensive intelligence, the lethal removal of terrorist opponents, the imprisonment of activists, or the development of strong defensive measures and structures. There has been much in the way of preventing or limiting the damage done by terrorist attacks. But some of Israel's tactical work has been counter-productive too. The 1982 Israeli invasion of Lebanon, for example, was successful in crushing the PLO there, but strategically painful in its generation of the hostile Hezbollah movement. Indeed, in Israeli counter-terrorist experience overall, there is a striking tension between tactical brilliance and strategic self-harm. International publicity has damaged Israel's standing; and some successes have simultaneously helped to sustain the terrorist resistance whose day-to-day expressions they have been aimed at combatting. Israel has tried simultaneously and understandably to achieve both strategic/partial strategic success and tactical victory; but it would be the former which allowed the terrorist threat to be more substantially addressed, and yet daily counter-terrorism— protecting life from attacks—has ironically made that larger goal more difficult to secure.

What about *inherent rewards*? Autonomous or partly autonomous of strategic, partial strategic, or tactical outcomes, such goods have repeatedly appeared in these pages. In the War on Terror, there was the catharsis of revenge against al-Qaida, Taliban, and ISIS enemies; there were certainly inherent rewards for some in the military, for some politicians, and for some businesses. There were painfully high costs for certain military people too, tragically offsetting the emotional and professional rewards that some experienced. In Northern Ireland, for many, revenge, camaraderie, and excitement accompanied the sense of the importance and inherent value of the counter-terrorist work pursued. Again, and importantly, there were also extremely painful aspects, with death, injury, trauma, and loss repeatedly emerging from the Troubles for those who worked for the state. Israel/ Palestine has seen the satisfaction for some of hitting back against terrorist enemies; there has also been political advantage for some figures and economic benefits for others.

A binary choice between yes or no is unlikely to prove persuasive in answer to the question *Does Counter-Terrorism Work?* Even the much-criticized post-9/11 War on Terror had considerable successes within it. My framework allows for a synoptic viewing of every aspect of counter-terrorist efficacy and inefficacy, for the discernment of wider patterns of behaviour, and for a more comprehensive analysis than would be attainable without it. For, emerging from the three case studies examined, but suggestive of wider intuitions about effective counter-terrorism more broadly, are the following wider-angled points. Having reflected systematically on counter-terrorist efficacy in long historical experience, what can we suggest about what will make counter-terrorism more likely to be effective in practice?

II

Throughout this book, the importance of *realistic goals, consistently pursued* has become very clear. Much of what proved problematic in Afghanistan, Iraq, and the wider War on Terror had its roots in a mixture of unrealistic ambitions and an unhelpful vacillation between priorities. In Northern Ireland, politicians and soldiers alike could at times make counter-terrorism more difficult by raising implausible hopes. Roy Mason, UK Secretary of State for Northern Ireland during the late 1970s, misleadingly promised to squeeze the PIRA like 'a tube of toothpaste';[9] influential British military figure Frank Kitson had suggested in 1971 that the conflict would 'be settled within five years'.[10]

Part of a more realistic approach probably involves learning to live with and to contain terrorism, rather than pledging to eradicate it (certainly within any short timeframe).[11] It also involves defining success at the partial strategic rather than at the strategic level and ensuring that tactical work supports rather than clashes with such higher-level goals. This also means avoiding an over-attention to tactical metrics (the number of kills, the number of prisoners, and so forth) and a resistance to the seduction of new technology (thinking, for instance, that sophisticated drones will solve counter-terrorist problems).

Repeatedly there has been a painful tension between the strategic and the tactical levels of counter-terrorism. This was perhaps most markedly evident in Israel/Palestine. But the War on Terror saw the difficulty that can arise when tactical success leads to an exaggeration of what can be achieved strategically and to a counter-productive over-confidence; in Afghanistan and Iraq, early tactical success by the military in the invasions of 2001 and 2003 respectively laid the foundations for disastrous miscalculation at the strategic level.

Setting and adhering to realistic goals also involves clarity regarding the nature of the terrorist threat itself and a resolve not to exaggerate nor panic. To exaggerate the novelty of what is faced at any given moment from terrorists risks amnesia about what is known to work in counter-terrorism and can lead to misjudged over-reaction. More broadly, inflating the danger posed by non-state terrorism leads away from life-saving containment and towards unrealistic approaches. Politicians have frequently slipped into this trap. After the appalling terrorist murder of a teacher in France in 2020, for example, President Emmanuel Macron described the state's fight against terrorism as 'existential'.[12] And it is not just politicians who do this. Others too have made this mistake: 'The greatest danger facing the world today comes from religiously inspired terrorist groups—often state sponsored— that are seeking to develop weapons of mass destruction against civilian targets.'[13] But this surely represents a distorting exaggeration; terrorist organizations 'can never pose an existential threat to a competent, well-governed state',[14] and the maintenance of a sense of proportion allows for a focus on sustaining normal life amid a residual, containable threat.

Historical and contemporary evidence suggests, in fact, that terrorism tends to represent a somewhat localized and limited threat.[15] Relevant too is the fact that 'there are relatively *few* people who engage in terrorism'.[16] Indeed, terrorism rarely embodies an existential threat to a state. Even where (as with 1970s West German terrorism) the violence is intended fundamentally to undermine the state and to attack the whole system associated with it,[17] the actual danger posed in practice can be comparatively trivial.

For victims, of course, terrorism has been an appalling and horrific threat. As societies, however, learning to live with it, and keeping it at the lowest possible levels, are mutually reinforcing aspects of the best approach. As shown in all three case studies, terrorism can be contained, despite the callous brutality of its agents and the awfulness of its human consequences. And, as evident also in those cases, some of the least effective counter-terrorism has emerged when states have exaggerated both the threat and the possibility of eliminating it. 'Terrorism operates'—as one of the world's foremost scholars of the subject has observed—'through subjective psycho-logical pressure. Its biggest facilitator is collective alarmism.'[18] Avoiding alarmism is therefore vital, and long-term containment, rather than eradica-tion, is the more fruitful and life-saving approach for states to adopt.[19] At times this has indeed been what Western states have come to do.[20] The more that phlegmatic calmness, a sense of appropriate proportion, and a culture of effective restraint are involved in counter-terrorism,[21] the more likely it is to work.

A second policy-facing point about effective counter-terrorism is a related one, since containment rather than extirpation will only be politically possible if societies are honest in recognizing *how difficult counter-terrorism is* in practice, and if they consequently facilitate consistently realistic, practical goal-setting. Terrorism is a career for which there exists no single psycho-logical profile;[22] terrorist groups are profoundly diverse and so are the individ-uals involved within them.[23] There is also the problem that counter-terrorist successes, when secured, are less visible than are those instances where the state has failed to prevent an attack: where efforts to prevent terrorism fail (and a bomb explodes), this tends to be more newsworthy and salient than when the police or the intelligence agencies stop an attack. 'MI5's operational successes are mostly invisible', as the organization's Director General Ken McCallum put it in 2020.[24] Again, in the words of former MI5 Director General Stella Rimington, 'When intelligence operations are successful and prevent a terrorist incident, no-one knows anything about it. It is very rare in those circumstances for anything to be said in public. The priority is to preserve the sources of intelligence. However, when intelligence fails to prevent an incident and a bomb does go off, there is a very high-profile disaster for all to see.'[25] Often, indeed, the ways in which counter-terrorist success has been achieved are such as to encourage the state to adopt a somewhat reticent approach: a well-placed and life-saving informer in a terrorist organization could become exposed if too much news is revealed and might therefore no longer be available in the future if all successes based on their intelligence work were to be made open.

Other significant difficulties also exist. Predicting trends in terrorist behaviour is challenging, owing to the great complexity of factors involved (and this book has stressed the importance of contingency in our understanding of counter-terrorism). Again, politicians and others in authority sometimes feel the need to focus on low-likelihood but catastrophic terrorism, for reasons which are understandable but with consequences that might lead attention away from more likely and immediate threats.[26]

We therefore require realistic goals, consistently pursued; and we need individuals and societies to recognize the difficulties and limitations of counter-terrorism in practice. Counter-terrorism will also be most effective if there is constant attention to the need for those involved to *avoid giving gifts to terrorists*. The risks here are well recognized and they include: the dangers of over-reliance on military methods; a failure to use intelligence appropriately; the degradation of proper legal approaches; a lack of effective coordination; and a disregard for the importance of credibility in counter-terrorist argument.[27]

Provoking over-reaction by the state is one of the longstanding aims of terrorist organizations.[28] State actors are as motivated by emotional human responses as other people, so this cycle of provocation and over-reaction is as explicable as it can be pernicious. Where states do avoid counter-productive over-reaction, counter-terrorist success has been more likely.[29] An over-reliance on military methods or an exaggeration of their utility in response to terrorism has repeatedly been problematic. Both terrorists[30] and counter-terrorists have repeatedly inflated the likely benefits and achievements to be secured through violent action. For states, the clumsy use of military force can give a huge gift to terrorist opponents. In the early phase of the Northern Ireland Troubles, British military heavy-handedness in Irish nationalist areas was often productive of terrorist sympathy. In the words of one early 1970s PIRA recruit, 'the British Army, the British government, were our best recruiting agents'.[31] Other first-hand testimony aligns with this. What were the reasons for joining the Irish republican armed struggle? They included 'the reaction of the security forces within the nationalist areas' during the early phase of the crisis;[32] in the words of another person who joined the PIRA, the behaviour of the British Army was 'a very, very important factor'.[33]

Repeated too has been the gift given to terrorists when states have failed to use counter-terrorism intelligence appropriately. In tactical-operational counter-terrorism, intelligence is foundational to success;[34] information (and its control) is absolutely crucial.[35] The gathering of information, the persuasion of people to provide information, the capacity to control the emergent narrative—all of these are highly significant. Indeed, the possession of information is possibly even more important in fighting terrorists than it is in orthodox war.[36]

As so often, there are echoes here between adversaries. For there is the long reality of terrorist groups themselves doing counter-intelligence work against their state opponents: gathering intelligence, penetrating state organizations, and having personnel in state bodies work for them.[37] For the state, understanding what terrorists want is essential. A failure to gather and interpret such information, or mistaking what such intelligence actually involves, can again give gifts to terrorist opponents. Intelligence-led policing has had repeated success against terrorist adversaries;[38] conversely, intelligence failures (starkly evident in different ways in relation to the Afghan and Iraqi experiences discussed earlier in this book) can be near-catastrophic. The importance of context has been stressed in preceding chapters. Without intelligence being gathered, interpreted, and properly utilized, an appreciation of context could prove elusive, and directly involved practitioners of various kinds have been eloquent on this point. 'Context is'—as David Omand points out—'needed to infer meaning'.[39] From a different perspective, UK diplomat and peace-maker Jonathan Powell has also been clear in his writings about the importance of intelligence: 'None of this is to say that terrorism can be dealt with in the absence of firm security policies and effective intelligence. Without the police successes against ETA, without the infiltration of the IRA by the intelligence agencies, and without the military campaign against the GAM, the conflicts in the Basque Country, Northern Ireland, and Aceh would not be over.'[40]

Given the disproportionately large effects produced by small numbers of terrorist actors and the fact that small-scale endeavours can have disproportionately large consequences in counter-terrorism too,[41] particularity and the uniqueness of context are absolutely crucial. Intelligence failures are therefore a major gift to terrorists, and so too is the transgressing of proper legal behaviour as this has an impact on crucial populations. The desire for an expanding range of data, and the increasing technological capacity to acquire it, can lead to tensions between security-oriented intelligence-gathering and individual rights of privacy.[42] Legal restraints and processes are crucial here, if counter-productive tactics are to be avoided; and the rule of law offers a strong foundation for the successful limitation of terrorism.[43] In democratic states, it is vital that responses to terrorism do not lead to a degradation in terms of civil liberties and other rights and freedoms.[44]

It has powerfully been argued that respect for human rights and for international law represents a crucial aspect of effective counter-terrorism.[45] It is unlikely that democratic states will respond to terrorism entirely according to either a criminal justice model or a military approach, since aspects of both are probably going to be deployed against sustained campaigns.[46] But, whatever the blend of approach, effective counter-terrorism in democracies will respect the equality of all citizens, it should involve guarantees of their rights,[47] and it should avoid the publicity own-goal likely to emerge if there is inadequate accountability and oversight. Nor is this a point necessarily opposed by counter-terrorists themselves. As former MI5 Director General Eliza Manningham-Buller conceives it, there exists no necessary tension between security and liberty: 'there is no liberty without security.'[48] States prepared to engage in full-scale repression and killing might indeed be able to crush terrorist organizations,[49] but even here the longer-term outcomes might be more complicated, and I take it as axiomatic that liberal democracies would rightly eschew such brutal responses.

A lack of effective coordination within and between states also offers a gift to terrorist enemies. Lack of cooperation and effective partnership between European states, for example, made the initial response to ISIS less successful,[50] and important practitioners have agreed with scholars' judgements that coordination is vital to effective counter-terrorism.[51] Where various wings of the state (and various states sharing a counter-terrorist cause) are not well coordinated, terrorist work is made significantly easier. As a consequence, counter-terrorism in many Western countries has now involved an attempt at organized coordination.[52]

Successes here are accompanied, of course, by some remaining challenges. The European Union's assessment nearly twenty years after 9/11 was that, in 2019, seven jihadist attacks were carried out within the EU, with twice that number being foiled by the authorities; overall during that year, there were 119 terrorist attacks in EU territories, with 1,004 arrests being made. These numbers reflected a diminishing rate of attacks (there had been 205 in 2017, and 129 in 2018).[53] They still reflected an enduring (though far from normality-shattering) challenge and the point here is that, without the coordination that has indeed been achieved, much more violence would almost certainly have occurred.

Another potential gift to terrorists is provided when states disregard the importance of credibility in counter-terrorist argument. For loyal citizens as for opponents or potential enemies, credible analyses of the issues involved in terrorism offer the basis both for shrewd policy and for being believed in these ongoing engagements. Of course, there are some who are unlikely to be persuaded even by the most evidence-based and cogently argued narratives. Conspiracy theorists such as those who claim that 9/11 was an inside operation by the US itself are engaged in pernicious political propaganda. This can make counter-terrorism work more difficult and it requires appropriate rebuttal,[54] together with the maintenance of credible arguments by those who seek to counter terrorist violence. Some minds will be closed, however credible the state's arguments. Nonetheless, most people are potentially open to persuasion, and the importance of states respecting the need for honest and credible narratives is therefore high if gifts are not to be given to terrorists.

Debates on radicalization (like terrorism, probably the wrong word, but one with which we now have to live) are relevant here. Counter-radicalization aims to deter or prevent people from being radicalized in the first place; deradicalization involves trying to draw those who have been radicalized away from their violent attitudes, thinking, ideology, and commitment. Both require patient, credible argument if they are to be effective. Disengagement is the process by which people are drawn away from their violent behaviours; but it need not mean that they cease to be supportive of their prior cause nor even that they have ceased activism entirely—they may now be peacefully pursuing the cause that was previously sought through violence. Crucial in this latter trajectory might be persuasion that violence is less likely to work than other forms of resistance, and so state credibility is clearly important here.

Opinions vary about the significance of ideology within terrorist motivation and campaigns[55] and it seems likely, for example, that the role of madrassas (Islamic religious schools) in generating jihadist terrorists has been exaggerated.[56] To the extent that ideological belief and conviction do play a part, however, it is a bonus to terrorists to allow such ideology to remain unchallenged where plausible refutation is possible. For example, there is clearly a terrorist-justificatory potential embodied in the interwoven arguments that Muslim lands will only flourish under sharia law, that Muslims are under assault from the West and from its apostate allies in Muslim countries, that only violent jihadist groups can defend against such attacks, and that those who oppose jihadist politics deserve punishment.[57] Counter-narratives here require appropriate timing, content, and audience-targeting, as well as the identification of the appropriate person to communicate the arguments. As with many terrorist accounts, what violent jihadists promulgate is often open to evidence-based and cogent counter-argument, and the disjunction between what most Muslims actually think and what groups like al-Qaida want them to think can either be reinforced or reduced by states' actions and approaches.

It is probably best not to see radicalization as something done to vulnerable, implicitly passive people. Rather, it tends to involve more complex relationships, and a set of active choices by those who radicalize.[58] Within this multicausal process, ideological disenchantment is only one element: disagreement with organizational decisions, actions, or policies can play a role; there can be personal antagonisms, disputes, disappointments, or a feeling of being overlooked by comrades within the movement; and there is sometimes a desire for money or respect. But it makes terrorist organizations' work easier if the states' arguments against movements lack credibility. Counter-narratives must resonate deeply, relevantly, and accessibly with the audiences which currently find terrorist arguments persuasive.[59] Misdiagnosing what is politically involved is therefore counter-productive.

Central in all this, as states try to counter terrorist organizations, is the normality of most terrorists. Martha Crenshaw argues in relation to terrorism that 'group dynamics matter more than individual motivations' and that:

> Groups using terrorism resemble other political organizations, so that over time organizational maintenance may supersede ideological purpose, rivalries among groups in competitive environments may drive behaviour, different roles and structures within organizations may matter to outcomes, and organizations may offer their members selective incentives, not just the pursuit of common ends.[60]

Painful though it might seem to accept the normality of those engaged in or supportive of callous terrorism, recognition of this point is vital if one is to avoid self-damaging approaches. Presenting terrorism as the realm of the insane makes counter-terrorism more difficult.

Accompanying the above themes is the next point: *use the gifts that terrorists give to the state*. The historical record demonstrates that, while strategic success for terrorists tends to be elusive, they frequently secure the objective of gaining publicity. But there is a paradox here. Terrorism repeatedly does seize publicity, but very often for acts which most people find repellent, including the harming of civilians and the vengeful attacking of the defence-less.[61] States can and should use this distinctive reality of terrorism, that its own tactical-publicity success can be turned into a victory for the state instead. Utilizing the paradox of terrorist publicity more systematically and subtly will tend to undermine sympathy for terrorist organizations and actions; and the movingly eloquent testimony of terrorism's victims[62] offers a powerful resource here in terms of clarifying what it is that terrorism most distinctively does in practice.

For public opinion is as vital to effective counter-terrorism as it is to terrorism itself. Large-scale protests in Spain (including those in the Basque Country) that were held to oppose ETA violence played a part in leading to that organization's eventual diminution.[63] Here, state partnership with allies (citizens, civil society, the media, the private sector) is important. If strategy is indeed 'the art of creating power',[64] then alliance-building to highlight the counter-productive brutality of terrorist violence offers one illustration. Terrorists' egregious transgression of people's human rights is a large-scale gift to state narratives, if properly utilized. When states harm or kill civilians in terrorism-related conflicts, it tends to strengthen support for terrorism.[65] So, too, in the other direction, with non-state atrocity.

A more complex point also emerging from our case studies, but relevant far beyond them, is the need *not to mistake the terrorist symptom for the more profound issues that are at stake*. In Israel/Palestine, in Northern Ireland, in Afghanistan, in Iraq, and in Syria it was issues of political legitimacy, of communal identity (and often nationalism), and of state power that were the deepest problems involved; appalling terroristic violence emerged from these, and required to be dealt with on that basis.[66]

Where counter-terrorism has most effectively dried up the reservoir of support for terrorist violence (such as in Northern Ireland), it is because it has been part of a wider political approach to the relevant root causes, an approach which does not exclusively focus on the terrorism. Counter-terrorist tactics have here facilitated at least partial strategic success as part of a wider body of state endeavour. Where counter-terrorism has been tactically successful but without decisive progress regarding the underlying conflict and politics (such as in Israel/Palestine), it has to be judged much less effective. Terrorism is a blood-stained symptom of the more important issues at stake. Historical understanding is crucial here, as a way of avoiding the solipsism of the present, and of avoiding the danger that contemporary violence will be reacted to in amnesiac and short-term fashion.

Counter-terrorism, if it is to be most effective, must recognize that issues such as the clash of rival nationalisms, the sustenance of certain societal power structures, state foreign policy, and international relations are all part of the work at hand. As one former Director General of MI5 observed, 'terrorism is resolved through politics and economics not through arms and intelligence, however important a role these play'.[67] The transformative politics of peace processes[68] can indeed be seen, therefore, as one arena within which effective counter-terrorist work can in practice be done.

If peace-making and peace-building are indeed integrated with counter-terrorism, then this has implications for state approaches to dialogue with terrorists. Politicians commonly claim that they will not negotiate with terrorists,[69] and the reasons are clear enough. There is the fear of giving prestige, credibility, and influence to violent non-state actors, of undermining non-violent advocates of the same cause, of seeming weak in the face of a terrorist threat, and of appearing insensitive to some victims' understandable hostility to any such contact with terrorists. But it is possible that successful use of dialogue can indeed be part of effective counter-terrorism, especially if one does accept that deeper issues are the cause of the violence.

During 1997–2007 Jonathan Powell was the chief British negotiator on Northern Ireland. He has powerfully argued for talking to terrorists more widely than just in that setting: 'when governments do engage with terrorists they almost always leave it far too late.... My experience in Northern Ireland convinced me that no conflict—however bloody, ancient, or difficult—is insoluble'; and 'it is always right to talk to terrorists, even if it may not always be the right moment to embark on a negotiation'.[70] If terrorism emerges from wider political and social conflicts and problems, if most of those who engage in and support it are as rational as other citizens, and if major terrorist groups do more than just violence and also have other political aspects to their work, then it can sometimes be the case that patient engagement might yield resolutions which embody partial strategic success in counter-terrorist terms. This is particularly likely if state tactical work can contain and limit terrorist efficacy, and if it becomes clear that terrorists' own strategic victory is not going to be secured by their further attacks.[71]

As noted above, there are dangers involved in such engagement by states, and the risks for politicians themselves (personally and in terms of political career) can be significant. Equally, as Tony Blair, Bill Clinton, and others discovered, there can be inherent rewards in terms of the positive benefits for one's political reputation if a peace process does help to end major terrorist violence. Significantly, when such outcomes are achieved, they can often occur (as in Northern Ireland) on the basis of the final deal falling far short of what the major terrorist adversary had long stipulated as essential for peace.

At heart here is the recognition that counter-terrorism is always political, and that it relates to issues of recognizably normal political concern. If al-Qaida have been motivated and self-justified significantly by Muslim perceptions of Western foreign policy, then that policy and narratives around it will form an important aspect of successful Western counter-terrorism. And counter-terrorism work often relates to normal, even banal, aspects of people's lives. There is strong evidence that locally informed, modest, securely implemented development programmes can aid violence-reduction in some settings that are experiencing terrorist attacks.[72]

Indeed, the most successful counter-terrorism is likely to relate to every part of politics, and to the kind of society (and state) that we choose to inhabit. This book is written by a political historian long intrigued by the Hobbesian problem of how to prevent people's enduringly rival views, rights, and interests from descending into violence. The great philosopher Thomas Hobbes rightly stressed that it is not satisfactory, when people hold divergent and simultaneously appealing views on a controversy, to expect resolution through one side being judged to possess right reason, effectively for one side to be judged to be right. More plausibly, he recognized, people on different sides will simultaneously and enduringly seek that their own opinion be judged the right one,[73] and will continue therefore to hold divergent views. My own argument is not that we should endorse Hobbes's proposed remedy, but rather: first, that we should recognize the decisive centrality of the problem that he identified (so persistently evident in Israel/Palestine, in Afghanistan, and in Northern Ireland, for example); second, that some political means do need to be found for preventing these clashes of rights and opinions from leading to violent conflict; third, that counter-terrorism forms one such political endeavour, and that effective counter-terrorism must therefore be conceived of in terms of the wider politics ultimately at stake in these situations. Counter-terrorism must be integrated into such broader politics, and failure to recognize this will lead repeatedly to failure. That is what has happened when Israeli tactics have obstructed strategic politics regarding conflict with Palestinians; that is what happened when the UK state allowed its military to behave clumsily in 1970s Northern

Ireland in the absence of political progress; that is what happened when the War on Terror witnessed a glib approach to politics and society (in Iraq, in Afghanistan) and was focused too much on military tactics and metrics.

As we reflect on our historical exploration of how far counter-terrorism has worked, some important intuitions have therefore emerged regarding what might make it most effective. There is the importance of setting and then consistently pursing realistic goals; it is vital that societies are honest about how difficult counter-terrorism is in practice, and why; states should not give gifts to terrorists (especially in the realms of the military, intelligence work, the law, coordination, and credibility); states should, however, use the gifts that terrorists present to them, especially in regard to terrorists' merciless and repellent violence and its human-rights effects on its victims; and it is crucial not to mistake the terrorist symptom for the more profound issues at stake in counter-terrorist endeavour: counter-terrorism must be seen as the political work that it necessarily is.

Another point is this: *states should think creatively about the resources that exist in society, such as those inherent in religious belief and practice, and those involved in civil resistance.*

While much attention has understandably been paid to the ways in which religion can fuel terrorist violence, it is also true that religious belief, commitment, and practice can offer powerful impulses towards the preventing, limiting, and restraining of terrorism.[74] This is evident, for example, in the teachings and example within Christianity towards compassion and empathy.[75] And it can be reinforced within Christianity by the tradition of people seeing their faith commitment as encouraging them towards appropriately benign social and political involvement and initiative.[76] Counter-terrorism in Western states tends away from a spiritual approach and towards broadly secular assumptions; on some occasions, indeed, a religious community can be presented as embodying a threat and as being central to the problem.[77] But it is very important for the countering of terrorism that people recognize the resources available among those of religious faith, since religion can be more of a peaceful than a violent force. It is vital here to note the contingent fluidity of religion, the numerous levels at which it simultaneously operates (global, institutional, group, and individual),[78] and the importance of ecumenical relationships of understanding between religious communities. Religion can work in many directions in relation to terrorism, and the state should utilize its positive potential and power. The Muslim Council of Britain emphatically condemned the violent attack on

Salman Rushdie in August 2022;[79] leading Jewish religious voices have argued strenuously that religion should not be used to justify terrorism, and indeed that religion offers a resource for opposing it;[80] in Northern Ireland, some reconciliatory developments which contributed to the limiting of terrorism have rested on friendships partly grounded in a shared Christian faith.[81] And the numbers associated with religious faiths are vast.[82]

Religion can therefore be a resource against terrorism, and this is especially important where people assume the opposite. Influential sources of terrorist inspiration often turn out, on serious investigation, to be undermined rather than strengthened by religious teaching. For example, in relation to its influence on Oklahoma bomber Timothy McVeigh, scholar Mark Juergensmeyer has referred to William Pierce's *The Turner Diaries* as possessing a 'strong Christian subtext'.[83] But this is rather misleading. Pierce's influential and repellent novel does carry very occasional and extremely vague references to God (made by the explicitly non-religious narrator Earl Turner),[84] but not to Christianity, whose distinctive teachings would clearly be opposed to the book's murderous brutality, to its lack of compassion and mercy, to its unempathic inhumanity, and to its appallingly divisive racial goals.[85] To the extent that Christian teaching relates to Pierce's book and to the ideas that are embodied within it, therefore, religious resources here decisively undermine the terrorist text.

Terrorism practised by people with some religious motivation has indeed been a phenomenon associated with numerous religious traditions, including Christianity, Islam, Buddhism, Judaism, and Hinduism.[86] But the peace-oriented teachings associated with those faith traditions are more important for states than is often realized in practice. Detailed understanding of religion is less profound among most people than would be useful here;[87] but greater positive engagement with this huge resource could have life-saving effects.

A not entirely unrelated source of counter-terrorist resource and influence is the increasingly recognized tradition of effective civil resistance. Central to strategic and partial strategic success in counter-terrorism is the undermining of longer-term support for violence, and of the belief that terrorism is justified because it is the only or the most effective way of securing political goals. There is significant scholarly agreement that non-state terrorism largely tends not to secure its practitioners' central political objectives.[88] Moreover, there is now a powerful body of scholarship complementing this point by stressing the greater efficacy of non-violent activism and civil resistance in relation to the achievement of political goals.[89]

If governments and other state actors are to persuade people not to engage in or not to persist in terrorism, then this argument surely offers a significant resource. If civil resistance offers more effective means of pursuing change and resisting injustice across many different settings, then the more pervasively and successfully this case is made, the better the situation might be for those seeking to reduce the incidence of terrorism. And this does sometimes intersect with the previous section concerning religion, especially where terrorism relates to politics with a religious dimension.[90] Central within terrorist campaigns is the issue of likely success.[91] If civil resistance is indeed more likely than terrorism to succeed, then stressing that reality can form part of the repertoire of effective, long-term counter-terrorism.

Counter-terrorism is most likely to succeed, *the more it can align with moral behaviour.* The damage done to counter-terrorist efforts by actions judged to lack morality has been evident in all three of our case studies, whether in terms of disregard for civilian deaths in the War on Terror or in Israel/Palestine, malign state/terrorist collusion in Northern Ireland, or the mistreatment of prisoners in all three settings. The immorality of terrorist atrocity itself is vitally important, and the moral choices involved in how we respond to terrorism are painful and profound.[92] The most effective counter-terrorism will be based on the maintenance of a distinction between what should be properly moral state action and the merciless violence of non-state terrorists. Some of this relates to complex questions regarding intelligence work. It is 'undeniable that the intelligence community, at least in Western democracies, wants to be seen as acting ethically—and not, it seems, purely for public relations reasons'.[93] Part of this relates to proportionality, part of it to legality, and part of it to transparent mechanisms of democratic oversight.[94] The scale and risks of counter-terrorist work must be in proportion to the scale and risks of the terrorist threats that they are aimed to counter.

A sophisticated case can be made that spying and intelligence work are indeed ethical if they can be judged to embody necessary, effective, and proportional ways of preventing violations of people's fundamental rights.[95] If counter-terrorism of this kind provides a means of protecting people which would not otherwise be achievable, and if the actions taken seem proportional to the good that is likely to be secured, then there is no reason to assume the immorality of this activity. Indeed, its avoidance might be deemed immoral. There are challenges in context, of course. It is clear that both the NSA in the US and GCHQ in the UK have engaged in mass surveillance, and both countries have introduced legislation to regulate these activities.[96] But concerns have long existed about the appropriateness of such state responses, with questions emerging about the invasion of privacy and about state misuse of powers that were established ostensibly for counter-terrorist purpose. There does exist real danger that state over-reaction to terrorism can unnecessarily damage civil liberties.[97]

Repeatedly, our analysis suggests that counter-terrorism has often proved less effective the less moral it has been. Over-aggressive soldiers, pernicious collusion with illegal groups, inattention to the risks of civilian harm, deception regarding justificatory arguments, the abuse of prisoners—all of these acts have proved counter-productive, and each of them is clearly unethical. The case of torture is particularly telling. Repeatedly damaging in terms of the standing and credibility of counter-terrorist states, it has been both less effective in generating new and actionable intelligence than other methods have proved, and also far more harmful to the cause of the state in question. Efficacy and ethics here align. There is a moral degradation of the state involved in the torturing of terrorist suspects, and the state's capacity to secure necessary intelligence, support, and sympathy for its cause is simultaneously harmed.[98]

In wider senses too, I think morality and integrity might be judged crucial to effective counter-terrorism. Nothing in Western treatment of Muslims seems to me to justify the kind of merciless and largely counter-productive terrorism that al-Qaida and ISIS sought to justify partly on those grounds.[99] But it seems unarguable that Muslims have experienced injustice within Western societies,[100] and to reduce such injustice is both inherently moral and also a fruitful way of countering justifications or reservoirs of support for jihadist terrorism.

At its heart, counter-terrorism is about protecting people from terrorist violence and from the terrible suffering and anguish that this violence can cause. This is in itself an ethically justifiable aim. The endemic suffering of so many people in terrorist conflicts raises important questions here, and the voices of the victims of violence still need to be more clearly heard than is often the case.[101] Callous, brutal behaviour has been at the heart of past terrorist campaigns, and profound damage has been done to so many victims of the violence[102] that the morality of counter-terrorism must be seen as foundational to it. This seems to me to demand that counter-terrorist operations should always seek to minimize human suffering rather than to treat human lives as unequal in their importance. I do not see why (for example) the life of a person who lives in the US or the UK should matter more than the life of a person who lives in Afghanistan or Iraq,[103] and so the issue of civilian deaths in terrorist/counter-terrorist conflict needs to be addressed accordingly as a priority.

In practice, of course, tensions will exist as moral choices are made. There are legitimate concerns that some counter-terrorist tactics (in surveillance, for example) might involve the transgression of people's rights, and on occasions the best outcome will probably be to opt for the lesser harm. It does not seem impossible that one might justify, in some cases, the infringing of privacy rights, if that infringement is considered a lesser evil than the harmful violence which such infringement prevents.[104]

Similarly, there exist clear moral tensions regarding intelligence work that includes the use of informers. The reasons behind people taking on the latter role are complex and varied, and some of these (greed, ego, a desire for revenge upon comrades with whom one has fallen out) are as unpleasant as are some of the aspects potentially involved in acquiring and running such agents (deceit, bribery, manipulation, coercion). It is also true, both that informer-based intelligence has saved many lives, and also that intelligence-led counter-terrorism has more to recommend it than some alternatives (the invasion and problematic occupation of other countries, for example). Moreover, for informers there can also be a combination of altruism, remorse, bravery, and redemption; the minimizing of human suffering represents a morally defensible act even if that work is done for partly distasteful reasons.

Nonetheless, for liberal democracies the practice of acquiring and utilizing secret intelligence remains morally challenging. Explicit rules, proper oversight, and genuine accountability for the intelligence community are important within a democratic society, not least because such structures help to define the kind of state that is sustained in the face of terroristic threats.

While it might be inappropriate to try to organize counter-terrorism too tightly in line with Just War thinking,[105] it does not seem unreasonable that counter-terrorist endeavour should echo some of the key approaches involved in such traditions. Counter-terrorist activity should involve sincere intention to pursue a justified cause (a cause such as prevention of the violent harm sought by terrorists); it should be carried out by appropriately delegated and accountably monitored actors; it should be judged necessary, and also proportionate to the good that it seeks to achieve as well as plausibly likely to advance that goal.[106]

Terrorists frequently see themselves to be virtuous, altruistic, moral, and righteous. Recognizing this is important if counter-terrorism is to be successful, and if counter-terrorist narratives are to be persuasive. A successful counter-terrorism will be a *just* counter-terrorism, and it needs to be widely seen as such.[107] If no persuasive analysis of counter-terrorism can ignore questions of morality,[108] then the comforting intuition one might draw from our case studies is that more moral counter-terrorism (well intentioned, life-saving, legally bound, honest, necessary, accountable, proportionate) tends to be more successful than less ethical state work.

Strengthening the state might seem a controversial element of counter-terrorist effectiveness but, for liberal democracies, it surely does play a part. The tactical-organizational aspects of this have been alluded to in our case studies, and the overarching theme of state power and capacity remains one of the key features of the dynamics involved in terrorism.[109] All layers of our framework are important here. Strategic or partial strategic success would involve the ending or substantial ending of the terrorist threat and—as in Northern Ireland, but as has proved elusive in Israel—this would embody a decisive strengthening of the state's resilience, legitimacy, and endurance. Part of this involves avoiding the demoralization or degradation of society, and not allowing society to be derailed too far from normality. State commitment and determination to outlast terrorist campaigns are vital to counter-terrorist success, so there needs to be a high-functioning and well-coordinated state, lastingly committed to opposing terrorists across all of its constituent parts,[110] and unwilling to be easily shifted by non-state terrorist violence.

At the tactical level too, day-to-day organizational strengthening is vital, both in terms of attack prevention and also a capacity to sustain normal life through recovery after terrorist incidents. Population control, another tactical aspect of counter-terrorism, involves the maintenance of state authority and power across the population, interwoven with the sustenance of order maintenance[111] as a means of preventing damage to health, limbs, or life. In all this, strong relationships of trust between those involved in counter-terrorism are important. Likewise, it is crucial that inherent rewards for those involved in counter-terrorism work to reinforce this state-strengthening process rather than clash with it. This is true in relation to politicians and officials, but also at symbolic and organizational level. In the 1970s, West German counter-terrorism was partly prioritized in order to enhance the international prestige, influence, standing, and salience of West Germany itself (a process which was marked by some success).[112] Again, the UK's proscription of terrorist organizations has direct effects in terms of how the state combats those groups. But arguably it also offers a broader, ritualized process through which to define the state itself as liberal, democratic, reasonable, accountable, and measured in contrast to the anti-democratic, extremist, illiberal terrorist enemy being proscribed.[113] Different cultures and norms of counter-terrorism persist in different states,[114] and these need to be ones which undermine terrorist opponents rather than damaging the state's own credibility and capacity (as on occasions occurred during the post-9/11 War on Terror).

The final link in this chain of policy-facing considerations regarding what will make counter-terrorism most effective involves *respect for the implications of long-term pasts and futures*. Historical-mindedness has been at the heart of this book, and counter-terrorism will suffer badly if it involves short-term memory, if it misunderstands historical periodization, and if it ignores contingent futures.

Repeatedly in this book we have seen the avoidable problems that have arisen when long-term historical dynamics have been overlooked. In the War on Terror, the actual causes behind al-Qaida's assault on the US were misdiagnosed, the historical inheritances of Afghanistan and Iraq were not appropriately respected, and the known knowns of effective counter-terrorism were overlooked (such as the dangers involved in over-militarization, in transgressing proper legal rules, in ignoring or misusing intelligence, in exaggerating the possibility of eradicating terrorism, or in proffering arguments that lacked credibility). In relation to Northern Ireland, pre-Troubles Irish experience had demonstrated the risks involved in mis-deploying the military in counter-terrorism, or in denying the political character of republican terrorist prison populations.[115] But—at great cost to counter-terrorist efforts—such pasts were insufficiently considered. Knowledge and insight that were available during the earliest days of the conflict were ignored, until painfully relearned at a later stage of the Northern Ireland Troubles. The authorized historian of MI5 (within the UK, a significant counter-terrorist agency) has noted that the absence of 'a long-term perspective' damaged the Security Service's early response to the Northern Ireland conflict, especially in terms of this amnesia leading to a repeat of the coordination problems that had been evident much earlier in relation to Irish terrorism; indeed, he suggests that the modern-day lack of policy-makers' attention to long-term trends has involved what he terms 'Historical Attention Span Deficit Disorder (HASDD)'.[116]

But historical long-termism can offer encouraging as well as depressing insights, since it offers potentially valuable guidance for contemporary response. Anarchism was arguably the dominant form of terrorism during the late nineteenth and early twentieth centuries. It had adherents in many countries, and during the late nineteenth and early twentieth centuries there were extensive efforts at international cooperation between states to counter it. These varied in their effectiveness, although they did achieve some striking and illuminating successes. Where police forces engaged in patient, well-resourced, careful, and coordinated intelligence work; where international cooperation, communication, and information-sharing between states operated efficiently; and where there was precise, professionalized protection-improvement for specific targets under threat—with all of this, anti-anarchist counter-terrorism became more effective. This was reinforced by a lack of popular sympathy for anarchist terrorism itself (another important historical detail). Moreover, where there was a heavy-handedly repressive approach by state authorities, counter-terrorism lost efficacy and could even prove counter-productive. Tellingly, in this anarchist period as so often, fear of terrorist conspiracy frequently exaggerated its actual threat.[117] In *all* of these aspects, and for all of the unquestionable differences between anarchism and latter-day jihadism, there were important foundations here on which to base a more life-protecting counter-terrorism in (for example) the War on Terror than actually obtained.

Historical-mindedness also involves nuanced rather than casual attitudes towards historical periodization. Some scholars have argued strongly for understanding terrorism's past in terms of successive waves.[118] Such an approach perhaps risks overlooking both the continuities between periods and also the extent to which historical experience moves forward jaggedly rather than in discrete periods of behaviour. The considerable differences between terrorist groups at any one historical moment, and the continuities across the dividing lines supposedly marking the emergence of each new wave, remain vital.

My own assessment is that a subtly flexible approach to historical periodization is best. This would recognize both continuity and change, acknowledging the interwovenness of past, present, and future; and it would allow for the retention rather than the abandonment of historically proven elements of what works best in counter-terrorism. Too rigid a set of supposed fault lines in the past probably obscures as much as it reveals. As noted earlier in this book, the idea that 9/11 necessarily represented a watershed offers a stark (and, in my view, unhelpful) example.[119] An exaggeration of how much had changed on that terrible day led to poor decisions by counter-terrorists.[120] More broadly, many shrewd scholars have resisted the idea that we should think too much in terms of new terrorism or of fault lines in terms of the fundamental dynamics of the phenomenon to be faced.[121] Of course, in some ways each new organization or each new phase carries some different aspects within it, as with any complex phenomenon. But the idea of a radically new terrorism, requiring utterly different responses, is probably mistaken.

Thinking historically about counter-terrorism therefore involves not only respect for actual pasts, and an avoidance of the dangers of Orwellian rectification of the past (that process of altering records of past events in order that those accounts match the current preferences and interests of the contemporaneously powerful).[122] It also involves analysing the relationship between change and continuity across long pasts, during which change occurs unevenly rather than in neat, epochal sequences.

Moreover, it involves respecting the contingency of the future. Given the heterogeneity of terrorism, and the complexity of each individual group and episode involved,[123] prediction is very difficult. Historians tend to be sceptical about predictive laws of human behaviour.[124] Indeed, if the future is read in terms of probabilities rather than firm predictions,[125] then this might align with historians' professional instincts. Such an approach will involve recognition both of the limits of predictive certainty, and also of the importance of those contextually based understandings which make some things seem more probable in the future than others. So, for example, it is hard to be sure of the changing dynamics of cyber threats in the future, though there seems little doubt about the current vulnerability of many states and societies to cyber attack, the need for far stronger defences to be established, and the implications that exist in the realm of counter-terrorism.[126] Undermining those cyber-communications networks used by terrorists would decisively damage them. There are certainly obstacles in the way of non-state terrorist engagement with cyberwarfare (not least in relation to skills capacity); but the possibilities for more extensive cyber-terrorism do exist,[127] so states need to think accordingly. As this unpredictable future unfolds, responsive agility will be required, in terms both of defence-strengthening and also of offensive capacity in order to compete in the cyber realm should that indeed become more of a focal point within which non-state adversaries work.

Central here is the admission of non-inevitability. Trajectories of terrorism and counter-terrorism very often prove to be contingent.[128] Just as in so many other areas of policy decision-making,[129] one has to deal with uncertainty when planning. Counter-terrorist teleologies (like terrorist ones) tend to be illusory, not least owing to the disproportionately large effect produced by the choices and behaviours of small numbers of people.[130] Contingency is absolutely central. And this is crucial as states approach the highest level of long-term endeavour, namely the attempt to end the conflicts of which terrorism is a grisly symptom. Jonathan Powell is absolutely right here: 'It is remarkable how quickly the shift can be from a conflict being "insoluble" to its solution being described as "inevitable" once an agreement is signed'; in truth, Powell suggests, 'Just as no conflict is insoluble, nor is it inevitable that it will be resolved at any particular moment in history. Believing that a solution is inevitable is nearly as dangerous as believing a conflict is insoluble.'[131]

These points all emerge from the systematically applied framework and the historical approach adopted in *Does Counter-Terrorism Work?* Indeed, I think that it is only through deploying a layered and comprehensive framework, and an historically minded attitude, that persuasive and policy-serious arguments are likely to emerge in this field.

This book has offered a detailed assessment of counter-terrorist efficacy in the past and, in doing so, it has been intended as a balanced historical account. But it contains within it also the hope that practical choices in the future might prove more and more capable of minimizing human suffering. Governments and others in power have heavy responsibility here, in terms of learning from the past and deriving practical wisdom from such reflection. But each of us has a duty to reflect on and contribute to how the future will respond to terrorism. This book's concluding argument has offered a series of historically grounded arguments about what is required: (1) we need realistic goals, consistently pursued; (2) societies must be honest in recognizing how difficult counter-terrorism is in practice; (3) states should avoid giving gifts to terrorists, and (4) they should use the gifts that terrorists give to the state; (5) we should not mistake the terrorist symptom for the more profound issues that are at stake; (6) states should think creatively about the resources that exist in society, such as those inherent in religious belief and practice, and those involved in civil resistance; (7) counter-terrorism is most likely to succeed the more it can align with moral behaviour; (8) we require a strengthening of the state, and also (9) a profound respect for the implications of long-term pasts and futures. My hope is that readers' engagement with such arguments, and with the case studies from which they grew, will make some contribution to our collective thinking about one of the world's most significant problems.

Notes

INTRODUCTION

1. My definition of terrorism is as follows: 'Terrorism involves heterogeneous violence used or threatened with a political aim; it can involve a variety of acts, of targets, and of actors; it possesses an important psychological dimension, producing terror or fear among a directly threatened group and also a wider implied audience in the hope of maximizing political communication and achievement; it embodies the exerting and implementing of power, and the attempted redressing of power relations; it represents a subspecies of warfare, and as such it can form part of a wider campaign of violent and non-violent attempts at political leverage' (R. English, *Terrorism: How to Respond* (Oxford: OUP, 2009), p. 24). Such a definition recognizes that states (as well as non-state actors) can practise terrorist violence. This book focuses on the efficacy of state attempts to counter non-state terrorism, and it therefore primarily assesses state action. It will also, however, reflect the extent to which state activity and non-state behaviour can exist in mutually shaping intimacy. For discussion of the definitional complexity surrounding the concept of the state itself, see R. English and C. Townshend (eds), *The State: Historical and Political Dimensions* (London: Routledge, 1999), pp. 2–8; for consideration of the state/non-state relationship, see the essays in R. English (ed.), *Illusions of Terrorism and Counter-Terrorism* (Oxford: OUP, 2015).

2. My own arguments concerning the efficacy of terrorism itself are set out systematically in R. English, *Does Terrorism Work? A History* (Oxford: OUP, 2016).

3. Ken McCallum, Speech to Journalists, 14 October 2020, https://www.mi5. gov.uk/news/director-general-ken-mccallum-makes-first-public-address.

4. M. Ranstorp, 'Mapping Terrorism Studies after 9/11: An Academic Field of Old Problems and New Prospects', in R. Jackson, M. Breen Smyth, and J. Gunning (eds), *Critical Terrorism Studies: A New Research Agenda* (London: Routledge, 2009), p. 22.

5. D. Muro (ed.), *When Does Terrorism Work?* (London: Routledge, 2018); M. Abrahms, *Rules for Rebels: The Science of Victory in Militant History* (Oxford: OUP, 2018); P. Krause, *Rebel Power: Why National Movements Compete, Fight, and Win* (Ithaca: Cornell University Press, 2017); English,

Does Terrorism Work?; P. R. Neumann and M. L. R. Smith, *The Strategy of Terrorism: How It Works, and Why It Fails* (London: Routledge, 2008).

6. M. Crenshaw, 'Constructing the Field of Terrorism', in E. Chenoweth, R. English, A Gofas, and S. N. Kalyvas (eds), *The Oxford Handbook of Terrorism* (Oxford: OUP, 2019), p. 719.

7. S. Pinker, *The Better Angels of Our Nature: The Decline of Violence in History and Its Causes* (London: Penguin, 2011), p. 344.

8. On the very significant theme of revenge as a motivation behind terrorism, see the excellent treatment in L. Richardson, *What Terrorists Want: Understanding the Terrorist Threat* (London: John Murray, 2006).

9. For those readers who would value a detailed analysis of the existing literature in the field, the book's Bibliographical Essay provides exactly that.

10. English, *Terrorism: How to Respond*, pp. 140–3.

11. 'State of the Union Address to the 107th Congress' (29 January 2002); 'Remarks on the Global War on Terror: The Enemy in Their Own Words' (5 September 2006); 'Address to the Nation on the Fifth Anniversary of 9/11' (11 September 2006) (*Selected Speeches of President George W. Bush 2001–2008* (Washington, DC: n.d.), pp. 103, 394, 426–7).

12. B. Ganor, *Israel's Counter-Terrorism Strategy: Origins to the Present* (New York: Columbia University Press, 2021), pp. 205, 299, 303.

13. English, *Does Terrorism Work?*, pp. 30–8.

14. In relation to counter-terrorism, I define strategy as the use of all available means to achieve specific political ends and policy objectives; connected with this, tactics operate at a lower level, involving the detailed and day-to-day choices involved in using the aforementioned means in line with strategic aims.

15. Repression and failure are two of the six categories for terrorist endings outlined in Audrey Cronin's important study; decapitation, negotiations, success, and reorientation are the others (A. K. Cronin, *How Terrorism Ends: Understanding the Decline and Demise of Terrorist Campaigns* (Princeton: Princeton University Press, 2009)).

16. C. Townshend, *Making the Peace: Public Order and Public Security in Modern Britain* (Oxford: OUP, 1993).

17. M. Crenshaw, *Explaining Terrorism: Causes, Processes, and Consequences* (London: Routledge, 2011), pp. 44, 125; J. Horgan, *The Psychology of Terrorism* (London: Routledge, 2005), pp. 50, 53, 62–5.

18. K. McConaghy, *Terrorism and the State: Intra-State Dynamics and the Response to Non-State Political Violence* (Basingstoke: Palgrave Macmillan, 2017).

19. D. Cannadine, *The Undivided Past: History beyond Our Differences* (London: Penguin, 2013).

20. Some people suggest a distinction between counter-terrorism and anti-terrorism. In practice, this often effectively involves defining the latter as a tactical sub-section of the former (S. D'Amato, *Cultures of Counter-Terrorism:*

French and Italian Responses to Terrorism after 9/11 (London: Routledge, 2019), pp. 14–15), and it seems to me that my layered framework can satisfactorily address both categories of engagement. For practically oriented reflections on defining counter-terrorism, see B. Ganor, *The Counter-Terrorism Puzzle: A Guide for Decision Makers* (London: Routledge, 2017; 1st edn 2005), pp. 25–46; see also D. Tucker and C. J. Lamb, *United States Special Operations Forces* (New York: Columbia University Press, 2007).

21. I refer to 'the past' as that which has happened before now; by 'history', I mean research and writing about the past; I use the term 'historiography' to refer to research and writing about history.

22. M. Crenshaw, 'Thoughts on Relating Terrorism to Historical Contexts', in M. Crenshaw (ed.), *Terrorism in Context* (University Park: Pennsylvania State University Press, 1995).

23. For further reflection on the distinctive contribution embodied in an historical approach to the understanding of terrorism, and also on historians' comparative absence from relevant debates to date, see English, *Does Terrorism Work?*, pp. 17–30; and R. English (ed.), *The Cambridge History of Terrorism* (Cambridge: CUP, 2021), pp. 4–18.

24. E. Hobsbawm, *On History* (London: Weidenfeld and Nicolson, 1997), p. 67.

25. See, for example, H. R. McMaster, *Battlegrounds: The Fight to Defend the Free World* (London: William Collins, 2020), pp. 1–2, 18–19, 425–31, 440, 442.

26. R. English, 'Change and Continuity across the 9/11 Fault Line: Rethinking Twenty-First-Century Responses to Terrorism', *Critical Studies on Terrorism*, 12/1 (2019).

27. It is suggestive, for example, that the number of terrorist attacks against France was significantly higher in the 1970s, 1980s, and 1990s than it has been since 2001 (D'Amato, *Cultures of Counter-Terrorism*, p. 66).

28. This aligns with how I think we should also study important related phenomena, such as nationalism (R. English, 'Directions in Historiography: History and Irish Nationalism', *Irish Historical Studies*, 37/147 (2011)).

29. See, for example, A. Silke, 'The Study of Terrorism and Counter-Terrorism', in A. Silke (ed.), *Routledge Handbook of Terrorism and Counter-Terrorism* (London: Routledge, 2019), p. 3.

30. My approach in this book therefore reinforces the arguments of those who seek fruitful dialogue between theoretical and empirical analyses; see S. Ashworth, C. R. Berry, and E. Bueno de Mesquita, *Theory and Credibility: Integrating Theoretical and Empirical Social Science* (Princeton: Princeton University Press, 2021), p. xi.

31. P. J. Bowler, *Progress Unchained: Ideas of Evolution, Human History, and the Future* (Cambridge: CUP, 2021).

32. J. Kay and M. King, *Radical Uncertainty: Decision-Making for an Unknowable Future* (London: Bridge Street Press, 2020), pp. xvi, 403, 424.

33. See, for example, R. J. Evans, *Altered Pasts: Counterfactuals in History* (London: Little, Brown, 2014); J. C. D. Clark, *Our Shadowed Present: Modernism, Postmodernism, and History* (London: Atlantic Books, 2003), pp. 28–9.

34. There is, of course, huge diversity in historians' approaches to their discipline, and considerable methodological change over time. Insightful essays on such themes are provided in M. Bentley, *Modern Historiography: An Introduction* (London: Routledge, 1999); and K. L. Klein, *From History to Theory* (Berkeley: University of California Press, 2012). But I still hold that my fivefold argument accurately outlines an enduringly distinctive historical approach.

35. J. Guldi and D. Armitage, *The History Manifesto* (Cambridge: CUP, 2014); English (ed.), *The Cambridge History of Terrorism*, pp. 18–22.

36. On aspects of what history can bring to the debate on terrorism and responses to terrorism, see the essays collected in English (ed.), *The Cambridge History of Terrorism*; on the regrettably ahistorical tendencies longstanding within the terrorism studies field, see A. Silke, 'Contemporary Terrorism Studies: Issues in Research', in Jackson, Breen Smyth, and Gunning (eds), *Critical Terrorism Studies*, pp. 45–6.

37. Cf. English, *Terrorism: How to Respond*.

38. English, *Does Terrorism Work?*; R. English, *Modern War: A Very Short Introduction* (Oxford: OUP, 2013); English (ed.), *Illusions of Terrorism and Counter-Terrorism*.

CHAPTER I

1. https://www.reaganfoundation.org/programs-events/webcasts-and-podcasts/podcasts/words-to-live-by/terrorism-1/.

2. D. C. Wills, *The First War on Terrorism: Counter-Terrorism Policy during the Reagan Administration* (Lanham: Rowman and Littlefield, 2003), pp. x, xii–xiii, 1–6, 10, 13–14, 22–44, 161, 213–16.

3. B. de Graaf, *Evaluating Counter-Terrorism Performance: A Comparative Study* (London: Routledge, 2013; 1st edn 2011), pp. 70–1, 76, 233.

4. See D. S. Chard, *Nixon's War at Home: The FBI, Leftist Guerrillas, and the Origins of Counter-Terrorism* (Chapel Hill: University of North Carolina Press, 2021), Nixon quotation at p. 168.

5. As so often in debates on terrorism, the phrasing is far from simple here. The War on Terror, the Global War on Terror, and the War on Terrorism have all been used. At times in recent years such phrasing has been less common than it was in the early period after 9/11. But there have existed many continuities across different US presidential regimes from 2001 until the present in terms of the fight against terrorism, and I think it is fair to refer to the post-9/11 period as one during which there has persisted a US-led War on Terror. That will be the focus of this chapter. Intriguingly, the Director General of the UK Security Service during 2002–7 did not think the War on Terror phraseology helpful (E. Manningham-Buller, *Securing Freedom* (London: Profile Books,

2012), p. 15). But, for all of its complexities and challenges, the US-led conflict after 9/11 could plausibly align with some definitions of war (English, *Modern War*, p. 16).

6. R. A. Clarke, *Against All Enemies: Inside America's War on Terror* (London: Free Press, 2004).

7. E. G. Arsenault, *How the Gloves Came Off: Lawyers, Policy-Makers, and Norms in the Debate on Torture* (New York: Columbia University Press, 2017), pp. 14, 85, 128–9, 167.

8. J. Comey, *A Higher Loyalty: Truth, Lies, and Leadership* (London: Macmillan, 2018), p. 81.

9. T. Blair, *A Journey* (London: Hutchinson, 2010), pp. 342, 345, 352, 369, 392.

10. Academic views vary here. Some scholars consider 9/11 to have inaugurated an unrecognizable, unprecedented, and new form of terrorism (Ganor, *The Counter-Terrorism Puzzle*, pp. xv, 273–5); I myself remain sceptical about such claims (English, 'Change and Continuity across the 9/11 Fault Line').

11. 'Address to the Nation on the September 11 Attacks' (11 September 2001), in *Selected Speeches of President George W. Bush 2001–2008*, pp. 57–8.

12. D. Kurtz-Phelan, 'Who Won the War on Terror?', *Foreign Affairs*, 100/5 (2021). Cf. J. Angstrom and I. Duyvesteyn (eds), *Understanding Victory and Defeat in Contemporary War* (London: Routledge, 2007).

13. 'Address to the Joint Session of the 107th Congress' (20 September 2001), in *Selected Speeches of President George W. Bush 2001–2008*, pp. 68, 70.

14. *National Strategy for Combating Terrorism* (September 2006), p. 3.

15. 'Address to the Joint Session of the 107th Congress' (20 September 2001), in *Selected Speeches of President George W. Bush 2001–2008*, p. 72.

16. 'Address to the Joint Session of the 107th Congress' (20 September 2001), in *Selected Speeches of President George W. Bush 2001–2008*, p. 68.

17. 'Address to the United Nations General Assembly' (10 November 2001), in *Selected Speeches of President George W. Bush 2001–2008*, pp. 88–9.

18. *National Strategy for Combating Terrorism* (September 2006), p. 1.

19. English, *Does Terrorism Work?*, pp. 51–8.

20. *Department of Homeland Security Strategic Framework for Countering Terrorism and Targeted Violence* (September 2019), p. 2.

21. *National Strategy for Combating Terrorism* (September 2006), pp. 1, 13.

22. 'State of the Union Address to the 107th Congress' (29 January 2002), in *Selected Speeches of President George W. Bush 2001–2008*, p. 103.

23. See, for example, J. Davis (ed.), *Africa and the War on Terrorism* (Aldershot: Ashgate, 2007).

24. C. Malkasian, *The American War in Afghanistan: A History* (Oxford: OUP, 2021), pp. 117–19, 130, 175, 240, 350–1, 396–8, 406–7, 443, 458–60.

25. T. H. Johnson, *Taliban Narratives: The Use and Power of Stories in the Afghanistan Conflict* (London: Hurst and Co., 2017), p. xxv.

26. 'Address to the Nation on Operations in Afghanistan' (7 October 2001), in *Selected Speeches of President George W. Bush 2001–2008*, pp. 75–6.

27. Quoted in Malkasian, *The American War in Afghanistan*, p. 113.

28. B. Hoffman and F. Reinares (eds), *The Evolution of the Global Terrorist Threat: From 9/11 to Osama bin Laden's Death* (New York: Columbia University Press, 2014).

29. S. Dorani, *America in Afghanistan: Foreign Policy and Decision Making from Bush to Obama to Trump* (London: I. B. Tauris, 2019), p. 2.

30. *Daily Telegraph*, 20 March 2021.

31. 'Statement by President Joe Biden on Afghanistan' (14 August 2021), https://www.whitehouse.gov/briefing-room/statements-releases/2021/08/14/statement-by-president-joe-biden-on-afghanistan/.

32. Kenneth McKenzie, quoted in *Guardian*, 31 August 2021.

33. M. Semple, 'The Unravelling of Afghanistan', *Irish Times*, 21 August 2021.

34. McMaster, *Battlegrounds*, pp. 188–90; Malkasian, *The American War in Afghanistan*, pp. 260, 297, 340–1, 451.

35. English, *Does Terrorism Work?*.

36. Tucker and Lamb, *United States Special Operations Forces*, pp. xiii, 7–17.

37. Malkasian, *The American War in Afghanistan*, pp. 201, 215, 224–5, 228, 262, 297–9, 304, 404, 434–5.

38. Malkasian, *The American War in Afghanistan*, p. 450.

39. B. G. Williams, *Counter Jihad: America's Military Experience in Afghanistan, Iraq, and Syria* (Philadelphia: University of Pennsylvania Press, 2017), p. 88.

40. On the pursuit and eventual killing of bin Laden, see P. Bergen, *Manhunt: From 9/11 to Abbottabad—The Ten-Year Search for Osama bin Laden* (London: Bodley Head, 2012).

41. Johnson, *Taliban Narratives*, p. 273.

42. English, *Terrorism: How to Respond*, pp. 127–31, 140–3.

43. Malkasian, *The American War in Afghanistan*, p. 111.

44. McMaster, *Battlegrounds*, pp. 169–70, 179.

45. S. G. Jones, *Counter-Insurgency in Afghanistan* (Santa Monica: RAND, 2008), pp. xii, 72, 116.

46. Malkasian, *The American War in Afghanistan*, pp. 96, 100–1, 131, 134, 144, 155, 215, 229, 329–30, 395, 456.

47. Malkasian, *The American War in Afghanistan*, pp. 92, 129–30.

48. Johnson, *Taliban Narratives*, p. xxvi.

49. T. Barfield, *Afghanistan: A Cultural and Political History* (Princeton: Princeton University Press, 2012; 1st edn 2010), pp. 141–2, 164, 172, 234–40, 255, 335.

50. Dorani, *America in Afghanistan*, p. 91.

51. S. Tankel, *With Us and Against Us: How America's Partners Help and Hinder the War on Terror* (New York: Columbia University Press, 2018), pp. 69, 73–4.

52. McMaster, *Battlegrounds*, p. 165; Williams, *Counter Jihad*, p. 95; Tankel, *With Us and Against Us*, p. 62; Barfield, *Afghanistan*, p. 333; Bergen, *Manhunt*, p. 49;

D. Kilcullen, *Blood Year: Islamic State and the Failures of the War on Terror* (London: Hurst and Co., 2016), p. 16.

53. S. Coll, *Directorate S: The CIA and America's Secret Wars in Afghanistan and Pakistan, 2001–2016* (London: Penguin, 2019; 1st edn 2018), p. 265; Jones, *Counter-Insurgency in Afghanistan*, p. 92; Dorani, *America in Afghanistan*, pp. 61–5, 81.

54. See the excellent treatment of these issues in Malkasian, *The American War in Afghanistan*.

55. See, for example, R. English, *Irish Freedom: The History of Nationalism in Ireland* (London: Pan Macmillan, 2006), pp. 11–19, 431–82.

56. Barfield, *Afghanistan*, pp. x, 7, 9, 18, 40–2, 91–2, 123, 317.

57. F. Christia, *Alliance Formation in Civil Wars* (Cambridge: CUP, 2012), p. 3.

58. See the superb treatment of this issue in Johnson, *Taliban Narratives*.

59. Johnson, *Taliban Narratives*, p. 284.

60. Quoted in English, *Does Terrorism Work?*, p. 55.

61. Williams, *Counter Jihad*, p. 27.

62. Barfield, *Afghanistan*, p. ix.

63. A. B. Zegart, *Spying Blind: The CIA, the FBI, and the Origins of 9/11* (Princeton: Princeton University Press, 2007), p. 41; cf. J. Burke, *The 9/11 Wars* (London: Penguin, 2011), pp. 33–4.

64. *The 9/11 Commission Report: Final Report of the National Commission on Terrorist Attacks upon the United States* (Washington, DC: US Government Printing Office, 2004), pp. 259–60, 263–7, 270–2, 277.

65. A. Lieven, *Pakistan: A Hard Country* (London: Penguin, 2012; 1st edn 2011), p. 478.

66. Jones, *Counter-Insurgency in Afghanistan*, p. 55.

67. McMaster, *Battlegrounds*, pp. 156–8.

68. D. Gillani, 'The History of Terrorism in Pakistan', in English (ed.), *The Cambridge History of Terrorism*.

69. Barfield, *Afghanistan*, p. 275.

70. J. Powell, *Talking to Terrorists: How to End Armed Conflicts* (London: Bodley Head, 2014), pp. 1, 345, 348–9, 352–3.

71. Christia, *Alliance Formation in Civil Wars*.

72. Malkasian, *The American War in Afghanistan*, pp. 79, 81, 86, 101, 155, 174, 199, 310–14, 426; cf. Barfield, *Afghanistan*, pp. 277, 313.

73. Even this, of course, is ambiguous. The July 2022 US killing of Ayman al-Zawahiri in Afghanistan reflected the possibility of ongoing Taliban/al-Qaida cosiness, although it also demonstrated American capacity to strike lethally in response (*Times*, 3 August 2022).

74. 'The manner of our withdrawal from Afghanistan was a disaster and a betrayal of our allies that will damage the UK's interests for years to come' (House of Commons Foreign Affairs Committee, *Missing in Action: UK Leadership and the Withdrawal from Afghanistan* (London: House of Commons, 2022), p. 3).

CHAPTER 2

1. Clarke, *Against All Enemies*, pp. 30–3, 241, 244, 265–7; Williams, *Counter Jihad*, pp. 75–6; C. Tripp, *A History of Iraq* (Cambridge: CUP, 2007), p. 272.
2. 'State of the Union Address to the 108th Congress' (28 January 2003), in *Selected Speeches of President George W. Bush 2001–2008*, p. 158.
3. P. Porter, *Blunder: Britain's War in Iraq* (Oxford: OUP, 2018), p. x.
4. R. N. Haass, *War of Necessity, War of Choice: A Memoir of Two Iraq Wars* (New York: Simon and Schuster, 2010; 1st edn 2009), p. 192.
5. Tripp, *A History of Iraq*, p. 277.
6. Clarke, *Against All Enemies*, p. xvi; Williams, *Counter Jihad*, pp. 23, 43–7, 138–41, 147, 149–53, 160.
7. Quoted in Richardson, *What Terrorists Want*, p. 232.
8. Williams, *Counter Jihad*, pp. 97, 107, 127, 129–32, 135–8, 160, 162–3, 166.
9. Quoted in J. Nixon, *Debriefing the President: The Interrogation of Saddam Hussein* (London: Bantam Press, 2016), p. 132.
10. English (ed.), *Illusions of Terrorism and Counter-Terrorism*.
11. E. Berman, J. H. Felter, and J. N. Shapiro, *Small Wars, Big Data: The Information Revolution in Modern Conflict* (Princeton: Princeton University Press, 2018), pp. 7, 23–4.
12. Clarke, *Against All Enemies*, pp. xiv, xvii, 273.
13. Clarke, *Against All Enemies*, p. xviii.
14. Richardson, *What Terrorists Want*, p. 234; M. Crenshaw and G. LaFree, *Countering Terrorism* (Washington, DC: Brookings Institution Press, 2017), p. 9; Davis (ed.), *Africa and the War on Terrorism*, p. 180.
15. Nixon, *Debriefing the President*, p. 4.
16. Manningham-Buller, *Securing Freedom*, p. 29.
17. Richardson, *What Terrorists Want*.
18. J. Zulaika, *Hellfire from Paradise Ranch: On the Frontlines of Drone Warfare* (Oakland: University of California Press, 2020), pp. 89–90; R. English, *Armed Struggle: The History of the IRA* (London: Pan Macmillan, 2012; 1st edn 2003).
19. English, *Does Terrorism Work?*, pp. 50–6.
20. T. K. Wilson, *Killing Strangers: How Political Violence Became Modern* (Oxford: OUP, 2020), pp. 88–9.
21. M. L. Cottam, J. W. Huseby, and B. Baltodano, *Confronting al-Qaida: The Sunni Awakening and American Strategy in al Anbar* (Lanham: Rowman and Littlefield, 2016).
22. Tripp, *A History of Iraq*, p. 275.
23. English, *Terrorism: How to Respond*, pp. 131–3.
24. Nixon, *Debriefing the President*, p. 35.
25. Porter, *Blunder*; Kilcullen, *Blood Year*, p. 10. The UK's Iraq War Inquiry pointed out that, 'Intelligence and assessments were used to prepare material to be used to support government statements in a way which conveyed

certainty without acknowledging the limitations of the intelligence' (*The Report of the Iraq Inquiry: Executive Summary* (6 July 2016), p. 46). Cf. F. Fukuyama, *After the Neocons: America at the Crossroads* (London: Profile, 2006), p. 6.

26. *Review of Intelligence on Weapons of Mass Destruction* (2004; Chairman: Lord Butler), pp. 150–2.

27. J. Waldron, *Torture, Terror, and Trade-Offs: Philosophy for the White House* (Oxford: OUP, 2010), p. 265; S. M. Hersh, *Chain of Command: The Road from 9/11 to Abu Ghraib* (New York: HarperCollins, 2004).

28. B. G. Williams, '*Aiqtihams* (Whirlwind Attacks): The Rise, Fall, and Phoenix-Like Resurgence of ISIS and Shiite Terrorists Groups in Iraq', in English (ed.), *The Cambridge History of Terrorism*.

29. Haass, *War of Necessity, War of Choice*, pp. 254–5.

30. Tripp, *A History of Iraq*.

31. C. Townshend, *When God Made Hell: The British Invasion of Mesopotamia and the Creation of Iraq, 1914–1921* (London: Faber and Faber, 2010).

32. McMaster, *Battlegrounds*, pp. 225, 242–3, 248, 253.

33. Cottam, Huseby, and Baltodano, *Confronting al-Qaida*, pp. 28–30, 35.

34. Tripp, *A History of Iraq*, pp. 282–3.

35. E. Sky, *The Unravelling: High Hopes and Missed Opportunities in Iraq* (London: Atlantic Books, 2015), pp. xi, 3–4, 11, 46.

36. Quoted in Tucker and Lamb, *United States Special Operations Forces*, p. 23.

37. J. Risen, *Pay Any Price: Greed, Power, and Endless War* (Boston: Houghton Mifflin Harcourt, 2014), pp. xv, 5, 11–13, 16–17, 19, 21–2, 24, 26–31.

38. N. Biggar, *In Defence of War* (Oxford: OUP, 2013), pp. 251–325.

39. Q. Cassam, *Vices of the Mind: From the Intellectual to the Political* (Oxford: OUP, 2019), pp. viii–x, 1–3, 5, 23–7.

40. J. Mueller and M. G. Stewart, *Chasing Ghosts: The Policing of Terrorism* (Oxford: OUP, 2016).

41. Quoted in S. Lindahl, *A Critical Theory of Counter-Terrorism: Ontology, Epistemology, and Normativity* (London: Routledge, 2018), p. 11.

42. Bergen, *Manhunt*, pp. 253–4; J. Mueller and M. G. Stewart, *Terror, Security, and Money: Balancing the Risks, Benefits, and Costs of Homeland Security* (Oxford: OUP, 2011), pp. 78, 165–6.

43. A. Mumford, *The West's War against Islamic State: Operation Inherent Resolve in Syria and Iraq* (London: I. B. Tauris, 2021).

44. Williams, *Counter Jihad*, p. xvi.

45. P. R. Neumann, *Bluster: Donald Trump's War on Terror* (London: Hurst and Co., 2019), pp. 94–5, 99–102, 104–5.

46. 'Twenty-Seventh Report of the UN Analytical Support and Sanctions Monitoring Team' (December 2020), p. 6.

47. 'Twenty-Seventh Report of the UN Analytical Support and Sanctions Monitoring Team' (December 2020), pp. 3, 5.

48. Mumford, *The West's War against Islamic State*.

49. President Barack Obama (10 September 2014), quoted in Mumford, *The West's War against Islamic State*, p. 25.

50. F. A. Gerges, *ISIS: A History* (Princeton: Princeton University Press, 2016). See also the valuable analyses in G. Wood, *The Way of the Strangers: Encounters with the Islamic State* (London: Penguin, 2018; 1st edn 2017); and A. Kruglova, *Terrorist Recruitment, Propaganda, and Branding: Selling Terror Online* (London: Routledge, 2023).

CHAPTER 3

1. Hoffman and Reinares (eds), *The Evolution of the Global Terrorist Threat*.

2. M. D. Silber, *The Al-Qaida Factor: Plots against the West* (Philadelphia: University of Pennsylvania Press, 2012).

3. Osama bin Laden (October 2001), quoted in Coll, *Directorate S*, p. 102.

4. Waldron, *Torture, Terror, and Trade-Offs*, p. 65.

5. Mueller and Stewart, *Chasing Ghosts*, pp. 267–73.

6. Zegart, *Spying Blind*, p. 196.

7. CONTEST: COuNter-TErrorism STrategy, presented to UK Cabinet and adopted, 2003; details not published until 2006; updated versions published in 2009, 2011, 2018.

8. *CONTEST: The United Kingdom's Strategy for Countering Terrorism* (June 2018), p. 7.

9. English, *Terrorism: How to Respond*, pp. 120–3.

10. D. Omand, *Securing the State* (London: Hurst and Co., 2010), pp. 79, 91.

11. *CONTEST* (2018), p. 30.

12. Manningham-Buller, *Securing Freedom*, p. 24.

13. *CONTEST* (2018), p. 8.

14. 'The Terrorism Act 2000 would prove to be the bedrock legislation for the domestic British war on terror after 9/11' (S. Hewitt, *The British War on Terror: Terrorism and Counter-Terrorism on the Home Front since 9/11* (London: Continuum, 2008), p. 35).

15. E. B. Tembo, *US-UK Counter-Terrorism after 9/11: A Qualitative Approach* (London: Routledge, 2014), p. 39.

16. Tankel, *With Us and Against Us*, p. 59.

17. D. Lowe, *Policing Terrorism: Research Studies into Police Counter-Terrorism Investigations* (London: CRC Press, 2019; 1st edn 2016), p. xxiii.

18. Lowe, *Policing Terrorism*, p. 77.

19. *Terrorism in Great Britain: The Statistics* (London: House of Commons Library, 2022; G. Allen, M. Burton, and A. Pratt), p. 7.

20. D. Omand, 'Foreword', in A. Staniforth and F. Sampson (eds), *The Routledge Companion to UK Counter-Terrorism* (London: Routledge, 2013), p. xxi.

21. Institute for Economics and Peace, *Global Terrorism Index 2022: Measuring the Impact of Terrorism* (Sydney: IEP, 2022), p. 2.

22. https://www.un.org/en/chronicle/article/losing-25000-hunger-every-day.

23. Cf. Omand, *Securing the State*, p. 89.

24. D'Amato, *Cultures of Counter-Terrorism*, pp. 101–2.

25. J. L. Gelvin, *The Modern Middle East: A History* (Oxford: OUP, 2016; 1st edn 2005), p. 358.

26. Richardson, *What Terrorists Want*, p. 2.

27. Eliza Manningham-Buller, Speech at Queen Mary's College London (9 November 2006), https://www.mi5.gov.uk/news/the-international-terrorist-threat-to-the-uk.

28. A. Brahimi, *Jihad and Just War in the War on Terror* (Oxford: OUP, 2010).

29. English, *Terrorism: How to Respond*, pp. 133–6.

30. Silber, *The Al-Qaida Factor*, p. 2.

31. N. Lahoud, 'Bin Laden's Catastrophic Success: Al-Qaida Changed the World—But Not in the Way It Expected', *Foreign Affairs*, 100/5 (2021); Davis (ed.), *Africa and the War on Terrorism*, pp. 170–1; H. A. Trinkunas, 'Financing Terrorism', in Chenoweth et al. (eds), *The Oxford Handbook of Terrorism*.

32. Twenty-Seventh Report of the UN Analytical Support and Sanctions Monitoring Team (December 2020), p. 6.

33. Manningham-Buller, *Securing Freedom*, p. 52.

34. *CONTEST* (2018), p. 13.

35. P. R. Neumann and R. Evans, 'Operation Crevice in London', in Hoffman and Reinares (eds), *The Evolution of the Global Terrorist Threat*, pp. 61–80.

36. L. Clutterbuck, 'Dhiren Barot and Operation Rhyme', in Hoffman and Reinares (eds), *The Evolution of the Global Terrorist Threat*, pp. 81–100. Cf. M. Innes and H. Innes, 'Counter-Terrorism Agencies and Their Work', in D. Muro and T. Wilson (eds), *Contemporary Terrorism Studies* (Oxford: OUP, 2022), p. 394.

37. Omand, *Securing the State*, p. 90.

38. J. N. Shapiro, *The Terrorist's Dilemma: Managing Violent Covert Organizations* (Princeton: Princeton University Press, 2013).

39. Zegart, *Spying Blind*, p. 68.

40. Mueller and Stewart, *Terror, Security, and Money*, pp. 1–4.

41. Mueller and Stewart, *Chasing Ghosts*, p. 67.

42. C. Lum, L. W. Kennedy, and A. Sherley, 'Are Counter-Terrorism Strategies Effective? The Results of the Campbell Systematic Review on Counter-Terrorism Evaluation Research', *Journal of Experimental Criminology*, 2/4 (2006).

43. Mueller and Stewart, *Terror, Security, and Money*.

44. Mueller and Stewart, *Terror, Security, and Money*.

45. Mueller and Stewart, *Chasing Ghosts*, pp. 7, 26–7.

46. R. English, 'History and the Study of Terrorism', in English (ed.), *The Cambridge History of Terrorism*, p. 9.

47. Staniforth and Sampson (eds), *The Routledge Companion to UK Counter-Terrorism*, p. 154.

48. C. McCauley, 'War versus Criminal Justice in Response to Terrorism: The Losing Logic of Torture', in W. G. K. Stritzke, S. Lewandowsky, D. Denemark, J. Clare, and F. Morgan (eds), *Terrorism and Torture: An Interdisciplinary Perspective* (Cambridge: CUP, 2009); de Graaf, *Evaluating Counter-Terrorism Performance*, pp. 171, 187–8; English, *Terrorism: How to Respond*, pp. 131–3; R. Crelinsten, *Counter-Terrorism* (Cambridge: Polity Press, 2009), p. 89.

49. B. W. Mobley, *Terrorism and Counter-Intelligence: How Terrorist Groups Elude Detection* (New York: Columbia University Press, 2012), pp. 244–7, 253; S. Hewitt, *Snitch! A History of the Modern Intelligence Informer* (London: Continuum, 2010); Omand, *Securing the State*, p. 12.

50. Zulaika, *Hellfire from Paradise Ranch*, p. 66.

51. Mobley, *Terrorism and Counter-Intelligence*.

52. For varying views, see R. Finegan, 'Targeted Killings: Perpetual War for Perpetual Peace?', in Silke (ed.), *Routledge Handbook of Terrorism and Counter-Terrorism*; M. M. Hafez and J. M. Hatfield, 'Do Targeted Assassinations Work? A Multivariate Analysis of Israel's Controversial Tactic during Al-Aqsa Uprising', *Studies in Conflict and Terrorism*, 29/4 (2006); B. C. Price, *Targeting Top Terrorists: Understanding Leadership Removal in Counter-Terrorism Strategy* (New York: Columbia University Press, 2019).

53. Price, *Targeting Top Terrorists*, pp. 29, 137–42, 145, 147, 176, 184–5, 187, 189.

54. M. J. Boyle, *The Drone Age: How Drone Technology Will Change War and Peace* (Oxford: OUP, 2020), pp. 6–7, 11, 60–3, 74–7, 82, 84, 90–1, 125, 277.

55. Tankel, *With Us and Against Us*, pp. 106, 153; F. Reinares, 'The 2004 Madrid Train Bombings', in Hoffman and Reinares (eds), *The Evolution of the Global Terrorist Threat*, p. 42.

56. Boyle, *Drone Age*, pp. 84–90.

57. Zulaika, *Hellfire from Paradise Ranch*, pp. 86–7, 90–1, 93–4.

58. J. Ralph, *America's War on Terror: The State of the 9/11 Exception from Bush to Obama* (Oxford: OUP, 2013), p. 53.

59. C. C. Fair, K. Kaltenthaler, and W. J. Miller, 'Pakistani Opposition to American Drone Strikes', *Political Science Quarterly*, 129/1 (2014).

60. Boyle, *Drone Age*, pp. 14–15, 24, 133–55, 161–5.

61. Birmingham Policy Commission, *The Security Impact of Drones: Challenges and Opportunities for the UK* (October 2014), pp. 6–7, 51, 61.

62. English, *Does Terrorism Work?*, pp. 204, 206–7, 246, 257.

63. See, for example, Burke, *The 9/11 Wars*, pp. 127–8, 252–6.

64. Hersh, *Chain of Command*.

65. Arsenault, *How the Gloves Came Off*, pp. 4–5, 134–5; Waldron, *Torture, Terror, and Trade-Offs*, pp. 7–8, 13–14, 16.

66. Stritzke et al. (eds), *Terrorism and Torture*, p. 2.

67. S. Zoller, *To Deter and Punish: Global Collaboration against Terrorism in the 1970s* (New York: Columbia University Press, 2021).

68. Ralph, *America's War on Terror*.

69. Coll, *Directorate S*, p. 27; see also the Report by S. Raphael, C. Black, and R. Blakeley, *CIA Torture Unredacted* (2019).

70. See, for example, S. O'Mara, *Why Torture Doesn't Work: The Neuroscience of Interrogation* (Cambridge, MA: Harvard University Press, 2015); J. W. Schiemann, *Does Torture Work?* (Oxford: OUP, 2016); A. H. Soufan and D. Freedman, *The Black Banners (Declassified): How Torture Derailed the War on Terror after 9/11* (London: Penguin, 2020; 1st edn 2011); R. E. Hassner, *Anatomy of Torture* (Ithaca: Cornell University Press, 2022); Nixon, *Debriefing the President*, pp. 60, 67, 85; Comey, *A Higher Loyalty*, p. 104.

71. Stritzke et al. (eds), *Terrorism and Torture*, pp. 2, 7, 9, 18, 26–7, 32–3, 63, 83; J. Zulaika, *Terrorism: The Self-Fulfilling Prophecy* (Chicago: University of Chicago Press, 2009), p. 215; Arsenault, *How the Gloves Came Off*, p. 174; L. Wright, *The Looming Tower: Al-Qaida's Road to 9/11* (London: Penguin, 2007; 1st edn 2006).

72. English, *Terrorism: How to Respond*, pp. 133–6.

73. Schiemann, *Does Torture Work?*, p. 36.

74. Arsenault, *How the Gloves Came Off*.

75. S. Scheipers, *Unlawful Combatants: A Genealogy of the Irregular Fighter* (Oxford: OUP, 2015), pp. 188–9, 194–202.

76. C. Gearty, *Liberty and Security* (Cambridge: Polity Press, 2013).

77. Richardson, *What Terrorists Want*.

78. The Uniting and Strengthening America by Providing Appropriate Tools Required to Intercept and Obstruct Terrorism (USA PATRIOT) Act of 2001 was introduced after 9/11 with a view to strengthening US national security. The Act extended the state's surveillance activity, aimed to improve coordination and communication between different agencies, and increased penalties for terrorist activity.

79. Comey, *A Higher Loyalty*, p. 151.

80. Lowe, *Policing Terrorism*, p. 67.

81. For reflections on aspects of the Snowden effect, see E. Bell, T. Owen, S. Khorana, and J. R. Henrichsen (eds), *Journalism after Snowden: The Future of the Free Press in the Surveillance State* (New York: Columbia University Press, 2017).

82. M. Parker, 'Tackling Terrorist Fundraising and Finances', in Silke (ed.), *Routledge Handbook of Terrorism and Counter-Terrorism*; Tankel, *With Us and Against Us*, pp. 98–9; Trinkunas, 'Financing Terrorism'.

83. *The 9/11 Commission Report*, pp. 339, 353.

84. G. F. Treverton, *Intelligence for an Age of Terror* (Cambridge: CUP, 2009).

85. Kevin McAleenan, Foreword, *Department of Homeland Security Strategic Framework for Countering Terrorism and Targeted Violence* (September 2019), cf. pp. 5, 12, 18.

86. *CONTEST* (2018), p. 7.

87. English, *Terrorism: How to Respond*, 136–40; Crenshaw and LaFree, *Countering Terrorism*, pp. 55, 61–3, 201; Jones, *Counter-Insurgency in Afghanistan*, p. 128; Omand, *Securing the State*, p. 14.

88. Zegart, *Spying Blind*, p. 78; D. Omand, *How Spies Think: Ten Lessons in Intelligence* (London: Viking, 2020), pp. 14–15, 208–32.

89. Omand, *How Spies Think*, p. 126.

90. Crenshaw, *Explaining Terrorism*, p. 14.

91. Coll, *Directorate S*, pp. 32, 36–9; Zegart, *Spying Blind*; A. B. Zegart, *Flawed by Design: The Evolution of the CIA, JCS, and NSC* (Stanford: Stanford University Press, 1999); Y. Veilleux-Lepage, *How Terror Evolves: The Emergence and Spread of Terrorist Techniques* (London: Rowman and Littlefield, 2020), pp. 5–6, 125–46, 151.

92. English and Townshend (eds), *The State*.

93. McMaster, *Battlegrounds*, p. 138; Boyle, *Drone Age*, pp. 72–3.

94. J. Grady, *Six Days of the Condor* (Harpenden: No Exit Press, 2015; 1st edn 1974), pp. 147–9, 152; S. Rimington, *At Risk* (London: Arrow Books, 2004), pp. 9–10, 56, 450–3.

95. Tankel, *With Us and Against Us*.

96. Price, *Targeting Top Terrorists*, p. 98.

97. Clarke, *Against All Enemies*, p. 286; Williams, *Counter Jihad*, p. 68.

98. Boyle, *Drone Age*, pp. 241–55.

99. M. J. Martin and C. W. Sassner, *Predator: The Remote-Control Air War over Iraq and Afghanistan: A Pilot's Story* (Minneapolis: Zenith Press, 2010), pp. 19, 31, 39, 42, 46–7, 51, 53–5, 61, 72, 85, 95, 100, 108, 128, 211–13, 215, 219, 235, 240, 258, 268, 283, 285–6, 290–1, 297, 299–300, 308, 310; Zulaika, *Hellfire from Paradise Ranch*, pp. 118, 122, 134.

100. Risen, *Pay Any Price*, p. 33.

101. B. S. Zellen, *State of Recovery: The Quest to Restore American Security after 9/11* (London: Bloomsbury, 2013).

102. Manningham-Buller, *Securing Freedom*, pp. 3, 5, 35, 46–7.

103. McMaster, *Battlegrounds*, pp. 3, 6; Soufan and Freedman, *The Black Banners (Declassified)*, pp. 159–60, 165, 171, 184; Zegart, *Spying Blind*, pp. 113–14, 124–5, 195.

104. Arsenault, *How the Gloves Came Off*, p. 152.

105. Quoted in Bergen, *Manhunt*, p. 21.

106. D. Byman, 'The Good Enough Doctrine: Learning to Live with Terrorism', *Foreign Affairs*, 100/5 (2021) .

107. Crenshaw, *Explaining Terrorism*, pp. 179, 181.

108. English, *Does Terrorism Work?*, pp. 63–72, 88–91; M. Abrahms, 'Al-Qaida's Scorecard: A Progress Report on Al-Qaida's Objectives', *Studies in Conflict and Terrorism*, 29/5 (2006).

109. English, *Terrorism: How to Respond*, pp. 127–31.

110. McMaster, *Battlegrounds*, p. 159.

111. R. English, 'The Future Study of Terrorism', *European Journal of International Security*, 1/2 (2016), pp. 144, 149.

112. Zegart, *Spying Blind*, pp. 69, 188.

113. On those complexities, see Gelvin, *The Modern Middle East*.

114. 'Address to the Nation on Operations in Afghanistan' (7 October 2001), in *Selected Speeches of President George W. Bush 2001–2008*, p. 77.

115. 'Address at the Citadel' (11 December 2001), in *Selected Speeches of President George W. Bush 2001–2008*, p. 94.

116. 'State of the Union Address to the 107th Congress' (29 January 2002), in *Selected Speeches of President George W. Bush 2001–2008*, p. 106.

117. Neumann, *Bluster*.

118. Neumann, *Bluster*, pp. 13, 18, 136–7, 143, 145, 150, 153–4, 157.

119. *National Strategy for Countering Domestic Terrorism* (June 2021), p. 5.

120. M. Fellman, *In the Name of God and Country: Reconsidering Terrorism in American History* (New Haven: Yale University Press, 2010).

121. A. Perliger, *American Zealots: Inside Right-Wing Domestic Terrorism* (New York: Columbia University Press, 2020).

122. English, 'Change and Continuity across the 9/11 Fault Line'.

123. Kevin McAleenan, Foreword, *Department of Homeland Security Strategic Framework for Countering Terrorism and Targeted Violence* (September 2019).

124. D. C. Rapoport, 'The Capitol Attack and the 5th Terrorism Wave', *Terrorism and Political Violence*, 33/5 (2021).

125. R. A. Pape and Chicago Project on Security and Threats, *Understanding American Domestic Terrorism: Mobilization Potential and Risk Factors of a New Threat Trajectory* (Chicago: University of Chicago, 2021).

126. *National Strategy for Countering Domestic Terrorism* (June 2021), p. 6.

127. B. F. Walter, *How Civil Wars Start: And How to Stop Them* (London: Viking, 2022), pp. ix–xii, xvii, 129–34, 157–8, 174.

128. Quoted in *Wall Street Journal*, 17 May 2022.

129. A. Macdonald [W. L. Pierce], *The Turner Diaries* (Laurel Bloomery: The National Alliance, 1978).

130. P. Gill, *Lone-Actor Terrorists: A Behavioural Analysis* (London: Routledge, 2015), p. 143.

131. Macdonald [Pierce], *The Turner Diaries*, quotations at pp. 1, 120, 145, 160.

132. J. M. Berger, *The Turner Legacy: The Storied Origins and Enduring Impact of White Nationalism's Deadly Bible* (The Hague: ICCT, 2016).

133. D. Dworkin, 'Terrorism: An American Story', in English (ed.), *The Cambridge History of Terrorism*.

CHAPTER 4

1. George Harrison, interviewed by the author, New York, 30 October 2000.

2. For different perspectives on the inheritances behind the Northern Ireland Troubles, see A. Jackson, *Ireland 1798–1998: War, Peace, and Beyond* (Chichester: Wiley-Blackwell, 2010; 1st edn 1999); English, *Irish Freedom*; B. O'Leary, *A Treatise on Northern Ireland. Volume 1—Colonialism: The Shackles of the State and*

Hereditary Animosities. Volume 2—Control: The Second Protestant Ascendancy and the Irish State. Volume 3—Consociation and Confederation: From Antagonism to Accommodation? (Oxford: OUP, 2019); P. Bew, *Ireland: The Politics of Enmity 1789–2006* (Oxford: OUP, 2007).

3. English, *Irish Freedom.*

4. N. J. Curtin, *The United Irishmen: Popular Politics in Ulster and Dublin 1791–1798* (Oxford: OUP, 1994); D. Dickson, D. Keogh, and K. Whelan (eds), *The United Irishmen: Republicanism, Radicalism, and Rebellion* (Dublin: Lilliput Press, 1993).

5. M. J. Kelly, *The Fenian Ideal and Irish Nationalism, 1882–1916* (Woodbridge: Boydell Press, 2006); N. Whelehan, *The Dynamiters: Irish Nationalism and Political Violence in the Wider World, 1867–1900* (Cambridge: CUP, 2012); J. Gantt, *Irish Terrorism in the Atlantic Community, 1865–1922* (Basingstoke: Palgrave Macmillan, 2010); D. G. Boyce, *Nineteenth-Century Ireland: The Search for Stability* (Dublin: Gill and Macmillan, 1990).

6. R. Wilson and I. Adams, *Special Branch—A History: 1883–2006* (London: Biteback, 2015); Whelehan, *The Dynamiters*; Curtin, *United Irishmen.*

7. C. Townshend, *The Republic: The Fight for Irish Independence* (London: Penguin, 2013); R. English, *Ernie O'Malley: IRA Intellectual* (Oxford: OUP, 1998); J. Augusteijn (ed.), *The Irish Revolution, 1913–1923* (Basingstoke: Palgrave Macmillan, 2002); D. G. Boyce (ed.), *The Revolution in Ireland, 1879–1923* (Basingstoke: Macmillan, 1988).

8. F. McGarry, *The Rising. Ireland: Easter 1916* (Oxford: OUP, 2010), p. 265.

9. 'Public Attitude and Opinion in Ireland as to the Recent Outbreak' (15 May 1916), Bonar Law Papers, House of Lords Record Office, London, BL 63/C/3.

10. English, *Ernie O'Malley*, pp. 81–2.

11. E. O'Malley, *On Another Man's Wound* (Dublin: Anvil Books, 1979; 1st edn 1936), p. 326.

12. M. Brennan, *The War in Clare 1911–1921: Personal Memoirs of the Irish War of Independence* (Dublin: Four Courts Press, 1980), pp. 80–1.

13. C. Townshend, *The Partition: Ireland Divided, 1885–1925* (London: Penguin, 2021), pp. 68, 80, 90; English, *Irish Freedom*, p. 317.

14. Townshend, *The Partition.*

15. English, *Armed Struggle*, pp. 145–7.

16. R. F. Foster, *Vivid Faces: The Revolutionary Generation in Ireland 1890–1923* (London: Penguin, 2014); F. Flanagan, *Remembering the Revolution: Dissent, Culture, and Nationalism in the Irish Free State* (Oxford: OUP, 2015); R. English, *Radicals and the Republic: Socialist Republicanism in the Irish Free State 1925–1937* (Oxford: OUP, 1994).

17. R. Alonso, *The IRA and Armed Struggle* (London: Routledge, 2007); English, *Armed Struggle*; H. McDonald and J. Holland, *INLA: Deadly Divisions* (Dublin: Poolbeg Press, 1994); E. Moloney, *A Secret History of the IRA* (London: Penguin, 2002); M. O'Doherty, *The Trouble with Guns: Republican Strategy*

and the Provisional IRA (Belfast: Blackstaff Press, 1998); H. Patterson, *The Politics of Illusion: A Political History of the IRA* (London: Serif, 1997; 1st edn 1989); J. Hepworth, 'The Age-Old Struggle': *Irish Republicanism from the Battle of the Bogside to the Belfast Agreement, 1969–1998* (Liverpool: Liverpool University Press, 2021).

18. S. Bruce, *The Red Hand: Protestant Paramilitaries in Northern Ireland* (Oxford: OUP, 1992); S. Bruce, 'The State and Pro-State Terrorism in Ireland', in English and Townshend (eds), *The State*; J. Cusack and H. McDonald, *UVF* (Dublin: Poolbeg Press, 2000; 1st edn 1997); A. Edwards, *UVF: Behind the Mask* (Newbridge: Merrion Press, 2017); I. S. Wood, *Crimes of Loyalty: A History of the UDA* (Edinburgh: Edinburgh Press, 2006); G. Mulvenna, *Tartan Gangs and Paramilitaries: The Loyalist Backlash* (Liverpool: Liverpool University Press, 2016).

19. J. D. Brewer and K. Magee, *Inside the RUC: Routine Policing in a Divided Society* (Oxford: OUP, 1991); E. Burke, *An Army of Tribes: British Army Cohesion, Deviancy, and Murder in Northern Ireland* (Liverpool: Liverpool University Press, 2018); C. Ryder, *The RUC 1922–2000: A Force Under Fire* (London: Arrow Books, 2000); A. Sanders and I. S. Wood, *Times of Troubles: Britain's War in Northern Ireland* (Edinburgh: Edinburgh University Press, 2012).

20. J. Powell, *Great Hatred, Little Room: Making Peace in Northern Ireland* (London: Bodley Head, 2008); G. J. Mitchell, *Making Peace* (London: William Heinemann, 1999); D. de Breadun, *The Far Side of Revenge: Making Peace in Northern Ireland* (Cork: Collins Press, 2008; 1st edn 2001); D. Godson, *Himself Alone: David Trimble and the Ordeal of Unionism* (London: Harper Perennial, 2005; 1st edn 2004); A. Sanders, *The Long Peace Process: The United States of America and Northern Ireland, 1960–2008* (Liverpool: Liverpool University Press, 2019).

21. For figures regarding responsibility for deaths in the conflict, see D. McKittrick, S. Kelters, B. Feeney, and C. Thornton, *Lost Lives: The Stories of the Men, Women, and Children Who Died as a Result of the Northern Ireland Troubles* (Edinburgh: Mainstream Publishing, 1999), p. 1495.

22. W. Matchett, *Secret Victory: The Intelligence War That Beat the IRA* (Lisburn: Hiskey Ltd, 2016); p. 8; for a similar argument, see M. Kirk-Smith and J. Dingley, 'Countering Terrorism in Northern Ireland: The Role of Intelligence', *Small Wars and Insurgencies*, 20/3–4 (2009), p. 566.

23. The consent principle requires that constitutional change for Northern Ireland can only occur if it has the support of a Northern Ireland majority.

24. A. McIntyre, *Good Friday: The Death of Irish Republicanism* (New York: Ausubo Press, 2008), pp. 7, 294.

25. Those that oppose the Good Friday Agreement (GFA) and the peace process.

26. Ken McCallum, Annual Threat Update, 14 July 2021, https://www.mi5.gov.uk/news/director-general-ken-mccallum-gives-annual-threat-update-2021.

27. M. J. Cunningham, *British Government Policy in Northern Ireland 1969–89: Its Nature and Execution* (Manchester: Manchester University Press, 1991), p. 252.

28. McKittrick et al., *Lost Lives*, p. 1494; https://cain.ulster.ac.uk/issues/violence/deathsfrom2002draft.htm.

29. Established after the 2015 Fresh Start Agreement between the UK and Irish governments and the Northern Ireland political parties, the IRC's role was to report on progress towards the ending of Northern Ireland paramilitarism.

30. Independent Reporting Commission, *Fourth Report* (7 December 2021), pp. 5, 9, 23.

31. J. Waller, *A Troubled Sleep: Risk and Resilience in Contemporary Northern Ireland* (Oxford: OUP, 2021), pp. 276–7.

32. https://www.mi5.gov.uk/threat-levels.

33. *Operation Banner: An Analysis of Military Operations in Northern Ireland* (London: Ministry of Defence, no date given), p. (II) 15.

34. M. Frampton, *Legion of the Rearguard: Dissident Irish Republicanism* (Dublin: Irish Academic Press, 2011); D. Reinisch, 'Teenagers and Young Adults in Dissident Irish Republicanism: A Case Study of Na Fianna Éireann in Dublin', *Critical Studies on Terrorism*, 13/4 (2020); J. F. Morrison, *The Origins and Rise of Dissident Irish Republicanism: The Role and Impact of Organizational Splits* (New York: Bloomsbury, 2013); J. Horgan, *Divided We Stand: The Strategy and Psychology of Ireland's Dissident Terrorists* (Oxford: OUP, 2013); P. M. Currie and M. Taylor (eds), *Dissident Irish Republicanism* (London: Continuum, 2011); S. A. Whiting, *Spoiling the Peace? The Threat of Dissident Republicans to Peace in Northern Ireland* (Manchester: Manchester University Press, 2015); A. Sanders, *Inside the IRA: Dissident Republicans and the War for Legitimacy* (Edinburgh: Edinburgh University Press, 2011); M. McGlinchey, *Unfinished Business: The Politics of 'Dissident' Irish Republicanism* (Manchester: Manchester University Press, 2019); A. Maillot, *Rebels in Government: Is Sinn Fein Ready for Power?* (Manchester: Manchester University Press, 2022), pp. 56–7.

35. R. English, 'Why Terrorist Campaigns Do Not End: The Case of Contemporary Dissident Irish Republicanism', in English (ed.), *Illusions of Terrorism and Counter-Terrorism*.

36. Hugh Orde, quoted in B. Rowan, *Political Purgatory: The Battle to Save Stormont and the Play for a New Ireland* (Newbridge: Merrion Press, 2021), p. 30.

37. 'Sinn Fein', Central Secretariat (Stormont Castle) Confidential Memo, 12 August 1985, Sinn Fein: Policy Group on Non-Violence Declarations 1985 File, PRONI CENT/1/14/16A.

38. SDLP Press Release, 26 September 1974, Linen Hall Library Political Collection (LHLPC) Archives, Belfast, SDLP Box 2.

39. *Irish Times*, 16 December 1971.

40. C. von Clausewitz, *On War* (London: Penguin, 1968; 1st edn 1832), p. 104.

41. Information Policy Coordinating Committee to the Secretary of State (31 January 1975), https://cain.ulster.ac.uk/proni/1975/proni_INF-2-15_1975-01-31_a.pdf.

42. F. Kitson, *Low Intensity Operations: Subversion, Insurgency, and Peacekeeping* (London: Faber and Faber, 2010; 1st edn 1971), pp. 2, 7, 201.

43. Cunningham, *British Government Policy in Northern Ireland 1969–89*, pp. 239–49.

44. M. J. Cunningham, *British Government Policy in Northern Ireland, 1969–2000* (Manchester: Manchester University Press, 2001), pp. 153–5.

45. *Operation Banner*, p. (IV) 1.

46. Blair, *A Journey*, p. 170; cf. Powell, *Great Hatred, Little Room*, p. 249.

47. Quoted in G. Spencer (ed.), *The British and Peace in Northern Ireland* (Cambridge: CUP, 2015), p. 299.

48. *Irish Times*, 4 November 1989.

49. Sanders and Wood, *Times of Troubles*, p. 249; N. Ó Dochartaigh, *Deniable Contact: Back-Channel Negotiation in Northern Ireland* (Oxford: OUP, 2021), p. 1; Cunningham, British Government Policy in Northern Ireland 1969–89, pp. 249–52; A. Cadwallader, *Lethal Allies: British Collusion in Ireland* (Cork: Mercier Press, 2013), pp. 361–2.

50. Sinn Fein, *Setting the Record Straight: A Record of Communications between Sinn Fein and the British Government October 1990–November 1993* (1994).

51. *Joint Declaration: Downing Street Declaration* (15 December 1993).

52. *The Framework Documents: A New Framework for Agreement* (22 February 1995).

53. *The Belfast Agreement: An Agreement Reached at the Multi-Party Talks on Northern Ireland* (April 1998), p. 2.

54. Spencer (ed.), *The British and Peace in Northern Ireland*, pp. 27–8, 31.

55. R. Montgomery, 'The Good Friday Agreement and a United Ireland', *Irish Studies in International Affairs*, 32/2 (2021), pp. 90–1.

56. *Anglo-Irish Agreement 1985: Agreement between the Government of Ireland and the Government of the United Kingdom* (15 November 1985), p. 3.

57. S. Kelly, *Margaret Thatcher, the Conservative Party, and the Northern Ireland Conflict, 1975–1990* (London: Bloomsbury, 2021). Thatcher's close friend and colleague Airey Neave was murdered by the INLA in London in March 1979; one member of the team which killed Neave was Patsy O'Hara, who subsequently died on hunger strike in 1981 (P. Bishop, *The Man Who Was Saturday: The Extraordinary Life of Airey Neave—Soldier, Escaper, Spymaster, Politician* (London: William Collins, 2019), p. 256). In 1984, the IRA came close to killing the Prime Minister herself, in an attack on the Conservative Party Conference in Brighton which left five people dead and many injured (English, *Armed Struggle*, p. 248).

58. P. J. McLoughlin, *John Hume and the Revision of Irish Nationalism* (Manchester: Manchester University Press, 2010); I. McAllister, *The Northern Ireland Social Democratic and Labour Party: Political Opposition in a Divided Society* (London: Macmillan, 1977); G. Murray and J. Tonge, *Sinn Fein and the SDLP: From Alienation to Participation* (Dublin: O'Brien Press, 2005).

59. G. Drower, *John Hume: Peacemaker* (London: Victor Gollancz, 1995), pp. 110–11; English, *Armed Struggle*, pp. 166, 240; *An Phoblacht/Republican News*, 2 February 1980.

60. *Irish Times*, 11 July 1972; *An Phoblacht/Republican News*, 22 November 1984; *An Phoblacht/Republican News*, 12 December 1985; G. Adams, *The Politics of Irish Freedom* (Dingle: Brandon, 1986), pp. 14, 126; Ó Dochartaigh, *Deniable Contact*, p. 62; English, *Armed Struggle*, pp. 339–40, 344, 348; R. W. White, *Out of the Ashes: An Oral History of the Provisional Irish Republican Movement* (Newbridge: Merrion Press, 2017), p. 97; P. Bishop and E. Mallie, *The Provisional IRA* (London: Corgi, 1988; 1st edn 1987), pp. 212–14.

61. PIRA statement, quoted in *Irish News*, 29 December 1969; *Republican News*, 23 August 1975; *An Phoblacht/Republican News*, 4 October 1980; PIRA spokesman, quoted in *An Phoblacht/Republican News*, 5 September 1981; PIRA spokesperson, quoted in *An Phoblacht/Republican News*, 5 January 1984; *An Phoblacht/Republican News*, 26 April 1984; Martin McGuinness, quoted in *An Phoblacht/Republican News*, 28 June 1984; PIRA spokesperson, quoted in *An Phoblacht/Republican News*, 18 October 1984; *An Phoblacht/Republican News*, 6 February 1986; English, *Armed Struggle*, p. 127; G. Adams, *A Pathway to Peace* (Cork: Mercier Press, 1988), p. 78; T. McKearney, *The Provisional IRA: From Insurrection to Parliament* (London: Pluto Press, 2011), p. ix.

62. Daithi O'Connell, quoted in *Irish Times*, 18 November 1974; piece by R. G. McAuley, Long Kesh, 14 March 1977, copy in Linen Hall Library Political Collection Archives, Belfast; *An Phoblacht/Republican News*, 14 February 1981; *An Phoblacht/Republican News*, 17 August 1989; *An Phoblacht/Republican News*, 14 February 1991; Martin McGuinness, quoted in Bishop and Mallie, *The Provisional IRA*, p. 227; English, *Armed Struggle*, p. 127; M. Kerr, *The Destructors: The Story of Northern Ireland's Lost Peace Process* (Dublin: Irish Academic Press, 2011), pp. 38–9, 270, 316, 320; Joe Cahill, quoted in T. Leahy, *The Intelligence War against the IRA* (Cambridge: CUP, 2020), p. 54. A distinction is sometimes drawn between calls for British withdrawal and for the withdrawal of British troops from Northern Ireland (e.g. Ó Dochartaigh, *Deniable Contact*, p. 55). But the PIRA did repeatedly call for British withdrawal itself (withdrawal of British sovereignty, of British rule) as a condition for ending their campaign. Moreover, in the context of 1970s civil war in the North, it is unimaginable that a withdrawal of British troops would have been seen as consistent with a long-term commitment to maintaining UK sovereignty.

63. M. McGuinness, *Bodenstown '86* (London: Wolfe Tone Society, n.d.), p. 3; McKearney, *The Provisional IRA*, p. 29; English, *Armed Struggle*, p. 125.

64. G. Bradley and B. Feeney, *Insider: Gerry Bradley's Life in the IRA* (Dublin: O'Brien Press, 2009), p. 123.

65. English, *Armed Struggle*, pp. 126–7.

66. R. W. White, *Ruairí Ó Brádaigh: The Life and Politics of an Irish Revolutionary* (Bloomington: Indiana University Press, 2006), pp. 174, 223, 226, 243, 246.

67. Bradley and Feeney, *Insider*, pp. 64, 66.

68. Quoted in White, *Ruairí Ó Brádaigh*, p. 243.

69. Sanders, *Inside the IRA*, pp. 54–5; English, *Does Terrorism Work?*, p. 103.

70. Jonathan Powell, in Spencer (ed.), *The British and Peace in Northern Ireland*, p. 315.

71. Ex-PIRA volunteer, interviewed by the author, Belfast, 26 November 2012.

72. 'H-Block Prisoners' Response to 1993 Downing Street Declaration', 21 February 1994, LHLPC Archives, Belfast.

73. Quoted in Kerr, *The Destructors*, p. 20.

74. D. Morrison, *Then the Walls Came Down: A Prison Journal* (Cork: Mercier Press, 1999), p. 97.

75. Quoted in Spencer (ed.), *The British and Peace in Northern Ireland*, p. 306.

76. Ó Dochartaigh, *Deniable Contact*.

77. R. J. Reed, 'Blood, Thunder and Rosettes: The Multiple Personalities of Paramilitary Loyalism between 1971 and 1988', *Irish Political Studies*, 26/1 (2011), 63; Kerr, *The Destructors*, pp. 256, 271, 307; Spencer (ed.), *The British and Peace in Northern Ireland*, p. 4; G. FitzGerald, 'The 1974–5 Threat of a British Withdrawal from Northern Ireland', *Irish Studies in International Affairs*, 17 (2006); A Guelke and F. Wright, 'The Option of a "British Withdrawal" from Northern Ireland: An Exploration of Its Meaning, Influence, and Feasibility', *Conflict Quarterly*, 10/4 (1990). This issue did not, of course, end with the 1970s: cf. 'Secretary of State's Meeting with No. 10 Policy Unit', 10 May 1984 memo re 9 May 1984 meeting, PRONI, Belfast, CENT/1/13/52A (File: Security and Prison Matters 31 October 1983–31 May 1984).

78. T. K. Whitaker, *A Note on North-South Policy* (11 November 1968), Lynch Papers, National Archives, Dublin, 2001/8/1; Lynch to Heath, 11 August 1970, National Archives, London, PREM 15/101; G. FitzGerald, *All in a Life: An Autobiography* (Dublin: Gill and Macmillan, 1992; 1st edn 1991), pp. 258–9; Kerr, *The Destructors*, pp. 253, 295.

79. Ó Dochartaigh's impressive book on negotiation offers no evidence that in the 1970s the PIRA thought it acceptable and deliverable to end their campaign on the basis which they later accepted, and which contained the consent principle and therefore no certainty about the ending of UK sovereignty (Ó Dochartaigh, *Deniable Contact*).

80. Ó Dochartaigh, *Deniable Contact*, pp. 143–5, 149.

81. R. Crelinsten, 'Conceptualizing Counter-Terrorism', in Silke (ed.), *Routledge Handbook of Terrorism and Counter-Terrorism*, p. 363.

82. Cunningham, *British Government Policy in Northern Ireland 1969–89*.

83. Among these were the UK's chief negotiator in the peace talks, Jonathan Powell; but figures such as Brendan Duddy, Michael Oatley, Robert McLaren, and Martin McGuinness had in various ways also played a crucial role in the long-term development of relationships, discussions, and possibilities. See the valuable treatment in Ó Dochartaigh, *Deniable Contact*.

84. Q. Thomas, 'Resolving Inter-Communal Conflict: Some Enabling Factors', in Spencer (ed.), *The British and Peace in Northern Ireland*, p. 110.

85. *Operation Banner*, pp. (I) 2–3.

86. Blair, *A Journey*, p. 160.

87. *Operation Banner*, p. (VIII) 3.

88. *Operation Banner*, p. (VIII) 15.

89. P. Dixon, 'Guns First, Talks Later: Neoconservatives and the Northern Ireland Peace Process', *Journal of Imperial and Commonwealth History*, 39/4 (2011); J. Bew and M. Frampton, '"Don't mention the war!": Debating Notion of a "Stalemate" in Northern Ireland (and a Response to Dr Paul Dixon)', *Journal of Imperial and Commonwealth History*, 40/2 (2012); P. Dixon, 'Was the IRA Defeated? Neo-Conservative Propaganda as History', *Journal of Imperial and Commonwealth History*, 40/2 (2012).

90. Morrison, *Then the Walls Came Down*, p. 71; Adams, *The Politics of Irish Freedom*, p. 58; IRA figure, quoted in T. P. Coogan, *The IRA* (London: Fontana, 1987; 1st edn 1970), p. 604; Laurence McKeown, quoted in Leahy, *The Intelligence War against the IRA*, p. 216; republican activists, quoted in White, *Out of the Ashes*, p. 342.

91. Hewitt, *Snitch!*, p. 66.

92. English, *Armed Struggle*, p. 285.

93. English, *Irish Freedom*.

94. Conor Murphy, interviewed by the author, Belfast, 16 December 2010.

95. Kelly, *The Fenian Ideal and Irish Nationalism*, pp. 2, 11, 47, 63–4, 132, 160, 239.

96. The biographer of one lastingly uncompromising Irish republican militant for example, observes that, 'Ruairí Ó Brádaigh is a complex man' (White, *Ruairí Ó Brádaigh*, p. xxii).

97. Quoted in Spencer (ed.), *The British and Peace in Northern Ireland*, p. 202.

98. P. Shirlow, J. Tonge, J. McAuley, and C. McGlynn, *Abandoning Historical Conflict? Former Political Prisoners and Reconciliation in Northern Ireland* (Manchester: Manchester University Press, 2010), pp. 98, 102.

99. *Operation Banner*, p. (I) 4.

100. During the years 1971–5, the Northern Ireland Troubles claimed the lives of 1,511 people (McKittrick et al., *Lost Lives*, p. 1494).

101. See, for example, the comments of Ulster loyalist Jackie McDonald, quoted in English, *Does Terrorism Work?*, pp. 226–7.

102. Shirlow et al., *Abandoning Historical Conflict?*.

103. E. Bergia, 'Unexpected Rewards of Political Violence: Republican Ex-Prisoners, Seductive Capital, and the Gendered Nature of Heroism', *Terrorism and Political Violence*, 33/7 (2021); Shirlow et al., *Abandoning Historical Conflict?*, p. 19.

CHAPTER 5

1. De Breadun, *Far Side of Revenge*, pp. 134, 412, 415–16.

2. G. McGladdery, *The Provisional IRA in England: The Bombing Campaign 1973–1997* (Dublin: Irish Academic Press, 2006).

3. Bradley and Feeney, *Insider*, pp. 151, 204. On the effects of infiltration, informers, and intelligence upon the PIRA's work, see also J. Crawley, *The Yank: My Life as a Former US Marine in the IRA* (Newbridge: Merrion Press, 2022), pp. 56, 83–4, 160, 163, 165, 206, 215, 231–5, 241–3, 249, 252, 260, 264.

4. Matchett, *Secret Victory*, p. 98.

5. *Operation Banner*, pp. (V) 1–2.

6. During the 1970s, 352 British Army personnel died in the Troubles; during the 1980s, the figure was 123 (McKittrick et al., *Lost Lives*, p. 1494).

7. E. Collins (with M. McGovern), *Killing Rage* (London: Granta, 1997), p. 152.

8. A. Edwards, *Agents of Influence: Britain's Secret Intelligence War against the IRA* (Newbridge: Merrion Press, 2021), p. 132.

9. Mobley, *Terrorism and Counter-Intelligence*, pp. 48–9. Cf. J. Holland and S. Phoenix, *Phoenix: Policing the Shadows. The Secret War against Terrorism in Northern Ireland* (London: Hodder and Stoughton, 1997; 1st edn 1996), p. 391; Bradley and Feeney, *Insider*, pp. 219–22, 234.

10. Matchett, *Secret Victory*, pp. 99–101.

11. S. Cursey, *MRF Shadow Troop* (London: Thistle Publishing, 2013), pp. 216–18.

12. W. Carlin, *Thatcher's Spy: My Life as an MI5 Agent inside Sinn Fein* (Newbridge: Merrion Press, 2021; 1st edn 2019).

13. A. Sanders, 'Northern Ireland: The Intelligence War 1969–1975', *British Journal of Politics and International Relations*, 3/2 (2011); T. Craig, '"You Will Be Responsible to the GOC": Stovepiping and the Problem of Divergent Intelligence Gathering Networks in Northern Ireland, 1969–1975', *Intelligence and National Security*, 33/2 (2018).

14. S. O'Callaghan, *The Informer* (London: Bantam, 1998); Moloney, *Secret History*, pp. 28–32, 155, 334–6; E. Moloney, *Voices from the Grave: Two Men's War in Ireland* (London: Faber and Faber, 2010), pp. 123, 276, 283–4; McIntyre, *Good Friday: The Death of Irish Republicanism*, pp. 179–95; Collins, *Killing Rage*; Holland and Phoenix, *Phoenix*; M. McGartland, *Fifty Dead Men Walking: The Heroic True Story of a British Secret Agent inside the IRA* (London: Blake, 1998; 1st edn 1997); M. Ingram and G. Harkin, *Stakeknife: Britain's Secret Agents in Ireland* (Dublin: O'Brien Press, 2004); McKearney, *The Provisional IRA*, p. 142.

15. Ex-RUC SB officer, interviewed by the author, County Down, Northern Ireland, 23 February 2010.

16. Matchett, *Secret Victory*, p. 112.

17. Ex-RUC HMSU officer, interviewed by the author, Belfast, 25 March 2010.

18. Ingram and Harkin, *Stakeknife*, p. 53; English, *Armed Struggle*, pp. 253–4.

19. Edwards, *Agents of Influence*, p. 154.

20. Holland and Phoenix, *Phoenix*.

21. C. Andrew, *The Defence of the Realm: The Authorized History of MI5* (London: Penguin, 2009), pp. 785, 795–7.

22. Cf. Police Ombudsman for Northern Ireland, *Statutory Report Relating to: Investigation into Police Handling of Certain Loyalist Paramilitary Murders and*

Attempted Murders in the North West of Northern Ireland during the Period 1989 to 1993 (Belfast: PONI, 2022), p. 88.

23. O'Callaghan, *The Informer*, pp. 26, 85, 89, 222, 228.

24. Collins, *Killing Rage*, quotation at p. 8.

25. Ingram and Harkin, *Stakeknife*, pp. 64–6.

26. Ingram and Harkin, *Stakeknife*; https://www.opkenova.co.uk/.

27. According to one former RUC officer, Donaldson had been told in December 2005 that he could be helped by police to leave Northern Ireland and escape to safety and a new life if he so chose, but that he decided not to do this (*Belfast Telegraph*, 30 March 2022).

28. K. McEvoy, *Paramilitary Imprisonment in Northern Ireland: Resistance, Management, and Release* (Oxford: OUP, 2001), p. 16.

29. Carlin, *Thatcher's Spy*, pp. 71, 152, 172, 214, 232.

30. English, *Armed Struggle*, p. 379.

31. Collins, *Killing Rage*, pp. 104–5.

32. M. J. McCleery, *Operation Demetrius and Its Aftermath: A New History of the Use of Internment without Trial in Northern Ireland, 1971–75* (Manchester: Manchester University Press, 2015).

33. Quoted in Sanders and Wood, *Times of Troubles*, p. 56.

34. English, *Armed Struggle*, pp. 148–54.

35. D. Omand and M. Phythian, *Principled Spying: The Ethics of Secret Intelligence* (Oxford: OUP, 2018), p. 7.

36. Police Ombudsman for Northern Ireland, *Statutory Report Relating to: Investigation into Police Handling of Certain Loyalist Paramilitary Murders and Attempted Murders in the North West of Northern Ireland during the Period 1989 to 1993* (Belfast: PONI, 2022), p. 91.

37. E. L. Williams and T. Leahy, 'The "Unforgiveable"? Irish Republican Army (IRA) Informers and Dealing with Northern Ireland Conflict Legacy, 1969–2021', *Intelligence and National Security*, published online 3 August 2022, https://www-tandfonline-com.queens.ezp1.qub.ac.uk/doi/full/10.1080/02684527.2022.2104000.

38. M. McGovern, 'Informers, Agents, and the Liberal Ideology of Collusion in Northern Ireland', *Critical Studies on Terrorism*, 9/2 (2016).

39. Cadwallader, *Lethal Allies*.

40. See the treatment in M. Smith, *UDR Declassified* (Newbridge: Merrion Press, 2022).

41. For a range of illuminating studies, see Bruce, *The Red Hand*; Bruce, 'The State and Pro-State Terrorism in Ireland', in English and Townshend (eds), *The State*; Cusack and McDonald, *UVF*; Edwards, *UVF*; Wood, *Crimes of Loyalty*; J. W. McAuley, *Very British Rebels? The Culture and Politics of Ulster Loyalism* (London: Bloomsbury, 2016); C. Crawford, *Inside the UDA: Volunteers and Violence* (London: Pluto Press, 2003); Mulvenna, *Tartan Gangs and Paramilitaries*; S. Nelson, *Ulster's Uncertain Defenders: Loyalists and the Northern Ireland Conflict* (Belfast: Appletree Press, 1984).

42. Edwards, *Agents of Influence*, p. 205; G. Gillespie, *Years of Darkness: The Troubles Remembered* (Dublin: Gill and Macmillan, 2008), pp. 92–5, 105–8, 112–15.

43. Edwards, *Agents of Influence*, p. 206; Edwards, *UVF*, pp. 159, 192, 216–18, 228–9, 304–5.

44. Ingram, 'Introduction', in Ingram and Harkin, *Stakeknife*, pp. 10, 12.

45. *The Report of the Patrick Finucane Review (The Rt Hon Sir Desmond de Silva QC)* (London: Stationery Office, 2012), p. (I) 6.

46. *The Report of the Patrick Finucane Review (The Rt Hon Sir Desmond de Silva QC)* (London: Stationery Office, 2012), pp. (I) 6–7.

47. *The Report of the Patrick Finucane Review (The Rt Hon Sir Desmond de Silva QC)* (London: Stationery Office, 2012), p. (I) 7.

48. *The Report of the Patrick Finucane Review (The Rt Hon Sir Desmond de Silva QC)* (London: Stationery Office, 2012), pp. (I) 12, 18–19.

49. J. O'Brien, *Killing Finucane: Murder in Defence of the Realm* (Dublin: Gill and Macmillan, 2005), pp. 11–12.

50. *The Report of the Patrick Finucane Review (The Rt Hon Sir Desmond de Silva QC)* (London: Stationery Office, 2012), p. (I) 11.

51. *The Report of the Patrick Finucane Review (The Rt Hon Sir Desmond de Silva QC)* (London: Stationery Office, 2012), pp. (I) 7–10, quotation at p. 8.

52. Police Ombudsman for Northern Ireland, *Statutory Report Relating to: Investigation into Police Handling of Certain Loyalist Paramilitary Murders and Attempted Murders in the North West of Northern Ireland during the Period 1989 to 1993* (Belfast: PONI, 2022), pp. 61, 102, 327.

53. Police Ombudsman for Northern Ireland, *Statutory Report Relating to: Investigation into Police Handling of Certain Loyalist Paramilitary Murders and Attempted Murders in the North West of Northern Ireland during the Period 1989 to 1993* (Belfast: PONI, 2022), pp. 90, 313–14, 321–2.

54. Leahy, *The Intelligence War against the IRA*, p. 141.

55. See the compelling account of this murder in I. Cobain, *Anatomy of a Killing: Life and Death on a Divided Island* (London: Granta, 2020).

56. D. B. Knox, *The Killing of Thomas Niedermayer* (Stillorgan: New Island Books, 2019), pp. 5–7, 72, 141, 158, 165, 170, 200, 213–16, 219–23, 271–5.

57. *Operation Banner*, p. (IV) 4.

58. Collins, *Killing Rage*, p. 152.

59. Leahy, *The Intelligence War against the IRA*.

60. G. A. Ackerman, 'The Provisional Irish Republican Army and the Development of Mortars', *Journal of Strategic Security*, 9/1 (2016); I. Woodford and M. L. R. Smith, 'The Political Economy of the Provos: Inside the Finances of the Provisional IRA—A Revision', *Studies in Conflict and Terrorism*, 41/3 (2018).

61. B. W. C. Bamford, 'The Role and Effectiveness of Intelligence in Northern Ireland', *Intelligence and National Security*, 20/4 (2005); J. Moran, 'Evaluating Special Branch and the Use of Informant Intelligence in Northern Ireland',

Intelligence and National Security, 25/1 (2010); J. Bew, M. Frampton, and I. Gurruchaga, *Talking to Terrorists: Making Peace in Northern Ireland and the Basque Country* (London: Hurst and Company, 2009).

62. *Operation Banner*, p. (VII) 6.
63. Gillespie, *Years of Darkness*, pp. 44–51, 58–61, 64–7, 71–4, 84–7.
64. English, *Does Terrorism Work?*, pp. 214, 255; Abrahms, *Rules for Rebels*.
65. *Irish Times*, 29 January 2022.
66. *Operation Banner*, p. (II) 7.
67. English, *Armed Struggle*, pp. 187–226.
68. 'Local Effects of the Hunger Strike', Central Secretariat (Stormont Castle) Confidential Memo, 17 August 1981, Hunger Strike File, PRONI CENT/1/10/66.
69. English, *Ernie O'Malley*, pp. 24–5.
70. 'Prisons: Offences in Prison', PRONI NIO/12/2A.
71. Telegram from M. Tierney to W. Whitelaw, received 15 June 1972, PRONI NIO/12/2A.
72. Statement by Public Relations Officer, Blanketmen, H-Blocks, 10 October 1980, LHLPC Archives, Belfast.
73. *H-Blocks: The Facts* (Published by the NIO, October 1980), PRONI NIO/2/1.
74. *H-Blocks: What the Papers Say* (Published by the NIO, July 1981) quoted press coverage favourable to the UK stance from newspapers in the UK, USA, Ireland, Canada, West Germany, France, India, Mauritius, New Zealand, Finland, Portugal, Switzerland, Sweden, Belgium, Spain, and Denmark (PRONI NIO/2/1).
75. English, *Armed Struggle*.
76. J. Stalker, *Stalker* (London: Harrap, 1988), pp. 9, 253; cf. R. W. White, T. Demirel-Pegg, and V. Lulla, 'Terrorism, Counter-Terrorism, and the "Rule of Law": State Repression and "Shoot-to-Kill" in Northern Ireland', *Irish Political Studies*, 36/2 (2021).
77. 'Repression of the Catholic Minority in Northern Ireland: List of Injuries Sustained by Remand Prisoners in Compound 6, Maze Prison, Long Kesh, on Friday 22 September 1972', PRONI D3564/2/5.
78. E. McGuire, *Enemy of the Empire: Life as an International Undercover IRA Activist* (Dublin: O'Brien Press, 2006), p. 43.
79. Kerr, *The Destructors*.
80. Gillespie, *Years of Darkness*, pp. 81–4, 95–8.
81. Ingram, in Ingram and Harkin, *Stakeknife*, p. 31.
82. *Report on the Interchange of Intelligence between Special Branch and CID, and on the RUC Units Involved, Including those in Crime Branch C1(1)* (Walker Report, 1980), quotation at p. 2.
83. *Belfast Telegraph*, 3 June 1991.
84. S. Rimington, *Open Secret: The Autobiography of the Former Director-General of MI5* (London: Hutchinson, 2001), pp. 117, 177.
85. Rimington, *Open Secret*, p. 179.
86. Martin Ingram, 'Introduction', in Ingram and Harkin, *Stakeknife*, p. 7.

87. Cursey, *MRF Shadow Troop*, pp. 53–4, 62, 66, 70, 72–3, 83–4, 89, 182, 203–4.

88. Carlin, *Thatcher's Spy*, pp. 30–2, 52, 76, 87, 165, 236, 239–40.

89. McGartland, *Fifty Dead Men Walking*, pp. 69, 71, 87, 122, 146, 173, 211, quotation at p. 83.

90. McEvoy, *Paramilitary Imprisonment in Northern Ireland*, p. 114.

91. Quoted in C. Ryder, *Inside the Maze: The Untold Story of the Northern Ireland Prison Service* (London: Methuen, 2000), p. 113.

92. McKittrick et al., *Lost Lives*, p. 1494.

93. Waller, *A Troubled Sleep*, p. 80.

94. See the reflections of John Garry at http://qpol.qub.ac.uk/public-opinion-ni-brexit-border/.

95. https://www.bbc.co.uk/news/politics/eu_referendum/results.

96. D. Cameron, *For the Record* (London: William Collins, 2019), p. 672.

97. D. Ferriter, *The Border: The Legacy of a Century of Anglo-Irish Politics* (London: Profile Books, 2019), pp. 129, 144.

98. English, *Armed Struggle*, p. 399.

99. Working Group on Unification Referendums on the Island of Ireland, *Interim Report* (London: The Constitution Unit, 2020), p. 42.

100. https://www.ark.ac.uk/nilt/2021/Political_Attitudes/REFUNIFY.html.

101. Institute of Irish Studies (University of Liverpool)/*Irish News, Opinion Poll 2022*, p. 7.

102. https://www.ark.ac.uk/nilt/2021/Political_Attitudes/REFUNIFY.html.

103. A. Reynolds, *My Autobiography* (London: Transworld, 2009), pp. 65–8, 204–14; English, *Armed Struggle*, p. 357.

104. W. D. Flackes and S. Elliott, *Northern Ireland: A Political Directory 1968–1993* (Belfast: Blackstaff Press, 1994; 1st edn 1980), pp. 359–412.

105. English, 'Why Terrorist Campaigns Do Not End', p. 137.

106. Maillot, *Rebels in Government*.

107. This is made more plausible, but by no means certain, by the increasing proportion of the Northern Irish population with a Catholic communal background (*Census 2021: Main Statistics for Northern Ireland. Statistical Bulletin—Religion* (22 September 2022)).

108. C. McCall, *Border Ireland: From Partition to Brexit* (London: Routledge, 2021).

109. Waller, *A Troubled Sleep*, pp. 268, 324.

110. English, *Terrorism: How to Respond*, pp. 118–43.

111. S. Rimington, *Secret Asset* (London: Arrow Books, 2006).

CHAPTER 6

1. J. M. Norman, 'Terrorism in Israel/Palestine', in English (ed.), *The Cambridge History of Terrorism*, p. 149.

2. See the evidence in the impressively rich study: B. Hoffman, *Anonymous Soldiers: The Struggle for Israel, 1917–1947* (New York: Alfred A. Knopf, 2015), pp. 97, 111–12, 143, 259, 300–1.

3. C. McCann, *Apeirogon: A Novel* (London: Bloomsbury, 2020).

4. See, for example, A. Hamilton, *Making Sense of the Bible: Rediscovering the Power of Scripture Today* (New York: HarperCollins, 2014), pp. 17–19, 25; B. M. Metzger and M. D. Coogan (eds), *The Oxford Companion to the Bible* (Oxford: OUP, 1993), pp. 349–52.

5. Gelvin, *The Modern Middle East*, pp. 191–204, 230.

6. English, *Irish Freedom*, pp. 12–21, 431–94.

7. L. Robson, *States of Separation: Transfer, Partition, and the Making of the Modern Middle East* (Oakland: University of California Press, 2017), p. 105.

8. Gelvin, *The Modern Middle East*, p. 116.

9. English, *Irish Freedom*, pp. 431–82.

10. On the struggle to establish Israel, see Hoffman, *Anonymous Soldiers*.

11. Hoffman, *Anonymous Soldiers*, p. 87.

12. P. Hayes (ed.), *Themes in Modern European History* (London: Routledge, 1992), p. 8.

13. Richardson, *What Terrorists Want*.

14. English (ed.), *Illusions of Terrorism and Counter-Terrorism*.

15. C. D. Freilich, 'Israel's Counter-Terrorism Policy: How Effective?', *Terrorism and Political Violence*, 29/2 (2017), p. 360.

16. The West Bank is referred to as Judea and Samaria by some Israelis and by the Israeli government.

17. C. Blattman, *Why We Fight: The Roots of War and the Paths to Peace* (New York: Viking, 2022), quotations at pp. 14–15, 57, 194.

18. See the very rich account in M. B. Oren, *Six Days of War: June 1967 and the Making of the Modern Middle East* (Oxford: OUP, 2002), pp. 1, 2, 8, 9, 13, 16, 17, 18, 21, 22, 24, 25, 31, 39, 42–3, 45, 49, 55, 63, 65, 66, 78–80, 82, 85–7, 92–3, 95, 97, 99, 102, 117–19, 143, 151, 159, 161, 313, 317–18.

19. Norman, 'Terrorism in Israel/Palestine', p. 160.

20. Veilleux-Lepage, *How Terror Evolves*.

21. Veilleux-Lepage, *How Terror Evolves*, p. 106.

22. English, *Does Terrorism Work?*, pp. 89–90, 206–7, 255; Abrahms, *Rules for Rebels*.

23. Named after Yitzhak Rabin (1922–95), Israeli Prime Minister during 1974–7 and 1992–5.

24. S. David, *Operation Thunderbolt: Flight 139 and the Raid on Entebbe Airport, the Most Audacious Hostage Rescue Mission in History* (London: Hodder and Stoughton, 2015).

25. A. R. Norton, *Hezbollah: A Short History* (Princeton: Princeton University Press, 2007), p. 33.

26. J. M. Norman, *The Second Palestinian Intifada: Civil Resistance* (London: Routledge, 2010).

27. B. Morris, *Righteous Victims: A History of the Zionist-Arab Conflict, 1881–1999* (New York: Alfred A. Knopf, 1999), p. 596.

28. G. Shafir, *A Half Century of Occupation: Israel, Palestine, and the World's Most Intractable Conflict* (Oakland: University of California Press, 2017), p. 19.

29. Mahmud al-Zahar, quoted in B. Hoffman, *Inside Terrorism* (New York: Columbia University Press, 2017; 1st edn 1998), p. 161.

30. See, for example, English, *Armed Struggle*, pp. 81–147.

31. S. Mishal and A. Sela, *The Palestinian Hamas: Vision, Violence, and Coexistence* (New York: Columbia University Press, 2006; 1st edn 2000).

32. R. Hassan, *Life as a Weapon: The Global Rise of Suicide Bombings* (London: Routledge, 2011), p. 77.

33. R. Singh, *Hamas and Suicide Terrorism: Multi-Causal and Multi-Level Approaches* (London: Routledge, 2011), p. 3.

34. Singh, *Hamas and Suicide Terrorism*, pp. 138–42.

35. English, *Does Terrorism Work?*, pp. 177–9; E. Berman, *Radical, Religious, and Violent: The New Economics of Terrorism* (Cambridge, MA: MIT Press, 2009), pp. 122, 129–32.

36. English, *Does Terrorism Work?*, pp. 148–85.

37. D. C. Kurtzer, S. B. Lasensky, W. B. Quandt, S. L. Spiegel, and S. Z. Telhami, *The Peace Puzzle: America's Quest for Arab-Israeli Peace, 1989–2011* (Ithaca: Cornell University Press, 2013).

38. *Declaration of Principles on Interim Self-Government Arrangements* (1993).

39. J. M. Skelly, 'Into the Labyrinth: Terrorism, History, and Diplomacy', in English (ed.), *The Cambridge History of Terrorism*.

40. Ganor, *Israel's Counter-Terrorism Strategy*, p. 181.

41. M. H. Yousef and R. Brackin, *Son of Hamas* (Milton Keynes: Authentic Media Limited, 2014; 1st edn 1999), pp. 127–32.

42. The al-Aqsa mosque is located at Haram al-Sharif.

43. D. Byman, *A High Price: The Triumphs and Failures of Israeli Counter-Terrorism* (Oxford: OUP, 2011), p. 8.

44. *Sharm El-Sheikh Fact-Finding Committee Report* ('Mitchell Report'), 30 April 2001.

45. Ganor, *Israel's Counter-Terrorism Strategy*, pp. 217, 226.

46. A. Merari, *Driven to Death: Psychological and Social Aspects of Suicide Terrorism* (Oxford: OUP, 2010), pp. 40, 271.

47. Ganor, *Israel's Counter-Terrorism Strategy*, p. 239.

48. Ganor, *Israel's Counter-Terrorism Strategy*, p. 247.

49. Berman, *Radical, Religious, and Violent*, p. 131.

50. Byman, *A High Price*, p. 110.

51. Ganor, *Israel's Counter-Terrorism Strategy*, p. 255; cf. Byman, *A High Price*, p. 202.

52. Byman, *A High Price*, p. 196.

53. Singh, *Hamas and Suicide Terrorism*, p. 68.

54. N. Qassem, *Hizbullah: The Story from Within* (London: Saqi, 2010; 1st edn 2005), pp. 9, 52, 54, 67–100, 135, 155.

55. Norton, *Hezbollah*, p. 77.
56. Norton, *Hezbollah*, pp. 101–12.
57. Norton, *Hezbollah*, p. 152.
58. In July 2022, for example, Israel shot down three Hezbollah drones that were heading towards an Israeli gas rig in the Mediterranean (https://www.bbc.co.uk/news/world-europe-62022452).
59. Ganor, *Israel's Counter-Terrorism Strategy*, p. 273.
60. Benjamin Netanyahu (20 December 2013), quoted in Ganor, *Israel's Counter-Terrorism Strategy*, p. 277.
61. Norman, 'Terrorism in Israel/Palestine', p. 168.
62. Freilich, 'Israel's Counter-Terrorism Policy', p. 365.
63. As with the August 2022 Israeli targeting of Palestinian Islamic Jihad with air strikes on Gaza, which saw senior PIJ figures (but also Palestinian children) killed (*Observer*, 7 August 2022; *Times*, 8 August 2022).

CHAPTER 7

1. Ganor, *The Counter-Terrorism Puzzle*, pp. 27, 39.
2. The three main organizations in Israel's intelligence community are: Shin Bet, the Israel Security Agency, the internal security service, with their Headquarters in Tel Aviv; Mossad, Israel's foreign intelligence service, with networks of informers, agents, and also friends globally; and Aman, Israeli military intelligence.
3. Ganor, *The Counter-Terrorism Puzzle*, p. 71.
4. Yousef and Brackin, *Son of Hamas*, pp. 67, 78–80, 98, 115, 118, 174.
5. David, *Operation Thunderbolt*.
6. Price, *Targeting Top Terrorists*, pp. 168–9.
7. Byman, *A High Price*, pp. 321, 323.
8. E. H. Kaplan, A. Mintz, S. Mishal, and C. Samban, 'What Happened to Suicide Bombings in Israel? Insights from a Terror Stock Model', *Studies in Conflict and Terrorism*, 28/3 (2005).
9. Price, *Targeting Top Terrorists*, p. 153.
10. G. Luft, 'The Logic of Israel's Targeted Killing', *Middle East Quarterly*, 10/1 (2003).
11. Zulaika, *Terrorism: The Self-Fulfilling Prophecy*, p. 11.
12. Shafir, *A Half Century of Occupation*, pp. 37–8.
13. Ganor, *Israel's Counter-Terrorism Strategy*, pp. 105–6.
14. P. Wilkinson, *Terrorism versus Democracy: The Liberal State Response* (London: Routledge, 2011; 1st edn 2001), pp. 172–3.
15. B. Netanyahu, *Fighting Terrorism: How Democracies Can Defeat Domestic and International Terrorists* (New York: Farrar Straus Giroux, 1995), p. 138; cf. English, *Terrorism: How to Respond*, pp. 131–3, 136–40.
16. A. Merari and S. Elad, *The International Dimension of Palestinian Terrorism* (Boulder: Westview Press, 1986), p. 90.

17. David, *Operation Thunderbolt*, p. 22.
18. Ganor, *Israel's Counter-Terrorism Strategy*; Ganor, *The Counter-Terrorism Puzzle*, pp. 153–5.
19. O. Hadad, 'A Battle of Names: Hamas and Israeli Operations in the Gaza Strip', *Terrorism and Political Violence*, 33/5 (2021).
20. Ganor, *The Counter-Terrorism Puzzle*, p. 111.
21. J. Gunning, *Hamas in Politics: Democracy, Religion, Violence* (London: Hurst and Co., 2009), pp. 42, 45, 95; B. Milton-Edwards and S. Farrell, *Hamas* (Cambridge: Polity Press, 2010), pp. vii–viii, 5, 156; M. Levitt, *Hamas: Politics, Charity, and Terrorism in the Service of Jihad* (New Haven: Yale University Press, 2006), pp. 5–6, 52, 97; R. Khalidi, *Palestinian Identity: The Construction of Modern National Consciousness* (New York: Columbia University Press, 2010; 1st edn 1997), p. 3.
22. Shafir, *A Half Century of Occupation*, pp. 62, 68.
23. A. Pedahzur, *The Israeli Secret Services and the Struggle against Terrorism* (New York: Columbia University Press, 2010; 1st edn 2009), pp. 8, 135.
24. Ganor, *Israel's Counter-Terrorism Strategy*, p. 190.
25. Byman, *A High Price*, p. 4; Zoller, *To Deter and Punish*, pp. 36, 57, 81.
26. https://hdr.undp.org/data-center/country-insights#/ranks; Freilich, 'Israel's Counter-Terrorism Policy', p. 368.
27. P. Roth, *Operation Shylock: A Confession* (London: Vintage, 2000; 1st edn 1993), p. 18.
28. Williams, *Counter Jihad*, p. 10.
29. English, *Does Terrorism Work?*, pp. 53–6.
30. Yitzhak Shamir, quoted in Ganor, *Israel's Counter-Terrorism Strategy*, p. 104.
31. Ganor, *Israel's Counter-Terrorism Strategy*, pp. 205, 308–9.
32. Byman, *A High Price*, p. 322.
33. C. Berrebi and E. F. Klor, 'On Terrorism and Electoral Outcomes: Theory and Evidence from the Israeli-Palestinian Conflict', *Journal of Conflict Resolution*, 50/6 (2006); C. Berrebi and E. F. Klor, 'Are Voters Sensitive to Terrorism? Direct Evidence from the Israeli Electorate', *American Political Science Review*, 102/3 (2008).
34. David, *Operation Thunderbolt*, pp. 321–2, 337, 342–3, 353–4, 371–2.
35. Ganor, *Israel's Counter-Terrorism Strategy*, p. 14. For reflections on the effects of emotion on state counter-terrorism, see McConaghy, *Terrorism and the State*.
36. Soldiers' testimonies, quoted in N. G. Finkelstein, *Gaza: An Inquest into Its Martyrdom* (Oakland: University of California Press, 2018), p. 80.
37. David, *Operation Thunderbolt*, pp. 226, 290–2, 294, 305–6, 309–10, 320–1, 323–4, 326, 340–1, 352–3.
38. Yousef and Brackin, *Son of Hamas*, pp. 134–5, 148, 151, 159, 162, 175, 178, 186, 207, 236, 258.
39. C. Berrebi and E. F. Klor, 'The Impact of Terrorism on the Defence Industry', *Economica*, 77/307 (2010).

40. Shafir, *A Half Century of Occupation*, pp. 46–7; Singh, *Hamas and Suicide Terrorism*.

41. M. Begin, 'From the Prime Minister's Office', in M. Begin, *The Revolt: Story of the Irgun* (Jerusalem: Steimatzky, 1977; 1st edn 1952), no page given.

42. Benjamin Netanyahu (18 December 2013), quoted in Ganor, *Israel's Counter-Terrorism Strategy*, p. 277.

43. Byman, *A High Price*, p. 368.

44. Ami Ayalon, quoted in Byman, *A High Price*, p. 365.

45. Pedahzur, *The Israeli Secret Services and the Struggle against Terrorism*, p. 3.

46. Byman, *A High Price*, pp. 5–6.

47. Byman, *A High Price*, pp. 4–5.

48. A. Pedahzur, L. Gutierrez, and A. Perliger, 'Israel: Can Terrorism Be Curbed?', in Silke (ed.), *Routledge Handbook of Terrorism and Counter-Terrorism*, p. 581. Cf. A. Pedahzur and A. Perliger, 'The Consequences of Counter-Terrorist Polities in Israel', in M. Crenshaw (ed.), *The Consequences of Counter-Terrorism* (New York: Russell Sage Foundation, 2010).

49. English, *Does Terrorism Work?*, pp. 182–3.

50. Richardson, *What Terrorists Want*, pp. 7, 148–9; Horgan, *The Psychology of Terrorism*, pp. 50, 53, 62–5.

51. I. Peleg and D. Waxman, *Israel's Palestinians: The Conflict Within* (Cambridge: CUP, 2011), pp. viii–ix, 7.

52. English, *Does Terrorism Work?*, pp. 183–4; Gunning, *Hamas in Politics*, pp. 2–3, 25, 56, 170–1, 236, 269; Milton-Edwards and Farrell, *Hamas*, p. 249.

53. Shafir, *A Half Century of Occupation*, p. 7.

54. Hoffman, *Anonymous Soldiers*.

55. Byman, *A High Price*, p. 2.

56. Ganor, *Israel's Counter-Terrorism Strategy*, pp. 7–10.

57. Ibrahim al-Quqa, quoted in Morris, *Righteous Victims*, p. 573.

58. N. Morag, 'Measuring Success in Coping with Terrorism: The Israeli Case', *Studies in Conflict and Terrorism*, 28/4 (2005), p. 318.

59. Freilich, 'Israel's Counter-Terrorism Policy'.

CHAPTER 8

1. 'Department of Defence Service of Remembrance at the Pentagon', in *Selected Speeches of President George W. Bush 2001–2008*, p. 79.

2. Quoted in B. Lawrence (ed.), *Messages to the World: The Statements of Osama bin Laden* (London: Verso, 2005), p. 111.

3. PRONI NIA1/3/9A. The memorial was unveiled at Parliament Buildings in Belfast in February 1984. Mr Graham was an Ulster Unionist Party politician and a law lecturer at Queen's University Belfast, where he was murdered by the PIRA in December 1983. The Euripidean quotation on the memorial plaque was chosen by his family.

4. G. Wilson, *Marie: A Story from Enniskillen* (London: Marshall Pickering, 1990), p. 2.

5. Quoted in Byman, *A High Price*, p. 1. Ben-Gurion was advising Ariel Sharon, at that time a special forces commander in Israel's army, and later to become Israeli Prime Minister himself.

6. English, *Does Terrorism Work?*, pp. 148-85.

7. Crenshaw and LaFree, *Countering Terrorism*, pp. 87-8; R. B. Jensen, *The Battle against Anarchist Terrorism: An International History, 1878-1934* (Cambridge: CUP, 2014), pp. 124-5, 341.

8. Mueller and Stewart, *Terror, Security, and Money*, pp. 147, 153.

9. Quoted in Moloney, *Secret History*, p. 149.

10. Kitson, *Low Intensity Operations*, p. 13.

11. English, *Terrorism: How to Respond*, pp. 120-3.

12. *Guardian*, 17 October 2020, https://www.theguardian.com/world/2020/oct/16/french-police-shoot-man-dead-after-knife-attack-near-paris-school.

13. A. M. Dershowitz, *Why Terrorism Works: Understanding the Threat, Responding to the Challenge* (New Haven: Yale University Press, 2002), p. 2; cf. S. Gorka, *Defeating Jihad: The Winnable War* (Washington, DC: Regnery Publishing, 2016), pp. 34, 112, 120, 123, 136.

14. Shapiro, *The Terrorist's Dilemma*, p. 270.

15. G. LaFree, L. Dugan, and E. Miller, *Putting Terrorism in Context: Lessons from the Global Terrorism Database* (London: Routledge, 2015), pp. 2-3, 126.

16. J. Horgan, *Walking Away from Terrorism: Accounts of Disengagement from Radical and Extremist Movements* (London: Routledge, 2009), p. 2.

17. G. Pridham, 'Terrorism and the State in West Germany during the 1970s: A Threat to Stability or a Case of Political Over-Reaction?', in J. Lodge (ed.), *Terrorism: A Challenge to the State* (Oxford: Martin Robertson, 1981).

18. C. Townshend, *Terrorism: A Very Short Introduction* (Oxford: OUP, 2018; 1st edn 2002), p. 16.

19. Richardson, *What Terrorists Want*, pp. 10-11.

20. *CONTEST: The United Kingdom's Strategy for Countering Terrorism* (June 2018), pp. 7-8, 13, 25, 30.

21. L. K. Donohue, *The Cost of Counter-Terrorism: Power, Politics, and Liberty* (Cambridge: CUP, 2008), pp. 336-8, 351-2.

22. Horgan, *Walking Away from Terrorism*, p. xxii.

23. Detailed research suggests that there is not a single profile even for lone-actor terrorists; it is true that a larger proportion of lone than of organizational actors have exhibited mental health problems, but how far those mental health problems explain the terrorism is another issue (Gill, *Lone-Actor Terrorists*, pp. 17-18, 28, 106-7, 120).

24. Ken McCallum, Speech to Journalists, 14 October 2020.

25. Rimington, *Open Secret*, p. 263; cf. Omand, *How Spies Think*, p. 6.

26. This is not to dismiss the importance of states trying to ensure the prevention of, for example, nuclear terrorism; serious-minded analyses of how best to do

exactly this have been produced, including M. Levi, *On Nuclear Terrorism* (Cambridge, MA: Harvard University Press, 2007); fictional evidence reflects the extent to which this nuclear danger has gripped very high-level political imaginations (H. R. Clinton and L. Penny, *State of Terror* (London: Pan Macmillan, 2021)).

27. English, *Terrorism: How to Respond*, pp. 127–43.
28. Crenshaw, *Explaining Terrorism*, p. 9; Richardson, *What Terrorists Want*.
29. Richardson, *What Terrorists Want*, p. 131.
30. English, *Does Terrorism Work?*
31. Tommy Gorman, interviewed by the author, Belfast, 2 May 2001.
32. Ex-PIRA volunteer, interviewed by the author, Belfast, 31 October 2001.
33. Tommy McKearney, interviewed by the author, Belfast, 20 September 2000.
34. J. Arquilla, *Bitskrieg: The New Challenge of Cyberwarfare* (Cambridge: Polity Press, 2021), p. 165.
35. Berman, Felter, and Shapiro, *Small Wars, Big Data*.
36. Arquilla, *Bitskrieg*, p. 22.
37. Hoffman, *Anonymous Soldiers*, p. 245.
38. Richardson, *What Terrorists Want*, pp. 12, 252. Cf. Omand, *How Spies Think*, pp. 109–10; Kitson, *Low Intensity Operations*, pp. 76–7, 168.
39. Omand, *How Spies Think*, p. 44.
40. Powell, *Talking to Terrorists*, p. 39.
41. Berman, Felter, and Shapiro, *Small Wars, Big Data*, pp. 10, 21, 266–8.
42. Boyle, *The Drone Age*, pp. 192–3.
43. S. Choi, 'Fighting Terrorism through the Rule of Law?', *Journal of Conflict Resolution*, 54/6 (2010).
44. See the especially important work of Conor Gearty here: C. Gearty, *Can Human Rights Survive?* (Cambridge: CUP, 2006); C. Gearty, *Civil Liberties* (Oxford: OUP, 2007); Gearty, *Liberty and Security*; cf. Waldron, *Torture, Terror, and Trade-Offs*, pp. 76–7.
45. T. Parker, *Avoiding the Terrorist Trap: Why Respect for Human Rights Is the Key to Defeating Terrorism* (London: World Scientific Publishing, 2019).
46. M. J. Boyle, 'The Military Approach to Counter-Terrorism', in Silke (ed.), *Routledge Handbook of Terrorism and Counter-Terrorism*.
47. P. T. Lenard, *How Should Democracies Fight Terrorism?* (Cambridge: Polity Press, 2020), p. 11. Cf. Crenshaw (ed.), *The Consequences of Counter-Terrorism*, p. 2.
48. Manningham-Buller, *Securing Freedom*, p. 36.
49. Hewitt, *The Effectiveness of Anti-Terrorist Policies*, p. 93.
50. P. R. Neumann, *Radicalized: New Jihadists and the Threat to the West* (London: I. B. Tauris, 2016), pp. xiii–xiv.
51. McMaster, *Battlegrounds*, pp. 279–80, 439, 442; Soufan and Freedman, *The Black Banners (Declassified)*, pp. 106–7, 160, 207, 244–5, 249–50, 288–92, 296–7, 300–2, 304–5, 333; cf. English, *Terrorism: How to Respond*, pp. 136–40.

52. European Commission, *A Counter-Terrorism Agenda for the EU: Anticipate, Prevent, Protect, Respond* (2020); *CONTEST: The United Kingdom's Strategy for Countering Terrorism* (June 2018).

53. European Union, *Terrorism Situation and Trend Report* (2020), pp. 5, 10–11.

54. Q. Cassam, *Conspiracy Theories* (Cambridge: Polity Press, 2019).

55. R. English, 'Terrorism and History: Current Knowledge and Future Research', in English (ed.), *The Cambridge History of Terrorism*, pp. 652–3.

56. A. Silke, 'Physical Facilitating Environments: Prisons and Madrassas as Mechanisms and Vehicles of Violent Radicalization?', in A. Richards, D. Margolin, and N. Scremin (eds), *Jihadist Terror: New Threats, New Responses* (London: I. B. Tauris, 2019), pp. 172–4.

57. K. Braddock, 'The Impact of Jihadist Terrorist Narratives and How to Counter Them: A Research Synthesis', in Richards et al. (eds), *Jihadist Terror*, p. 15.

58. K. McDonald, *Radicalization* (Cambridge: Polity Press, 2018).

59. See the superb treatment of these themes in Q. Cassam, *Extremism: A Philosophical Analysis* (London: Routledge, 2022).

60. Crenshaw, *Explaining Terrorism*, p. x.

61. Abrahms, *Rules for Rebels*; Berman, Felter, and Shapiro, *Small Wars, Big Data*.

62. T. Knatchbull, *From a Clear Blue Sky: Surviving the Mountbatten Bomb* (London: Hutchinson, 2009); J. D. Parker, *On the Waterfront* (Durham: Pentland Press, 2000); D. Harris-Gershon, *What Do You Buy the Children of the Terrorist Who Tried to Kill Your Wife? A Memoir* (London: Oneworld Publications, 2013).

63. R. Alonso, 'Why Do Terrorists Stop? Analyzing Why ETA Members Abandon or Continue with Terrorism', *Studies in Conflict and Terrorism*, 34/9 (2011).

64. L. Freedman, *Strategy: A History* (Oxford: OUP, 2013), pp. xii, 607.

65. Berman, Felter, and Shapiro, *Small Wars, Big Data*, pp. 189, 192, 196–200.

66. English, 'The Future Study of Terrorism'. For an important set of relevant reflections, see A. Abu-Bakare, 'Counter-Terrorism and Race', *International Politics Reviews*, 8 (2020).

67. Manningham-Buller, *Securing Freedom*, p. 15.

68. J. D. Brewer, *Advanced Introduction to the Sociology of Peace Processes* (Cheltenham: Edward Elgar, 2022).

69. Skelly, 'Into the Labyrinth', pp. 594–5.

70. Powell, *Talking to Terrorists*, pp. 2, 4, 41. See also I. Taylor, *The Ethics of Counter-Terrorism* (London: Routledge, 2018).

71. English, *Does Terrorism Work?*

72. Berman, Felter, and Shapiro, *Small Wars, Big Data*, pp. 148, 150–1, 157–8, 183, 221, 258, 260, 321.

73. T. Hobbes, *Leviathan* (London: Fontana, 1962; 1st edn 1651), p. 82; R. Tuck, *Hobbes* (Oxford: OUP, 1989), pp. 57–8; R. Tuck, *The Rights of War and Peace: Political Thought and the International Order from Grotius to Kant* (Oxford: OUP, 1999), pp. 131–2; N. Malcolm, *Aspects of Hobbes* (Oxford: OUP, 2002), pp. 31, 44.

74. G. Ganiel, 'Praying for Paisley—Fr Gerry Reynolds and the Role of Prayer in Faith-Based Peacebuilding: A Preliminary Theoretical Framework', *Irish Political Studies*, 36/1 (2021); R. English, 'Terrorism, History, and Religion'.

75. See, for example, Matthew 5:7, 7:12, 20:34, 22:39; Mark 6:34, 8:2, 12:31; Luke 6:31, 6:36, 7:13, 15:20.

76. D. English, *Christianity and Politics* (Belfast: Queen's University Belfast, 1993); L. G. Jones, *Christian Social Innovation: Renewing Wesleyan Witness* (Nashville: Abingdon Press, 2016).

77. For sophisticated consideration of one British version of this trend, see S. Croft, *Securitizing Islam: Identity and the Search for Security* (Cambridge: CUP, 2012).

78. J. Mayall and S. Silvestri, *The Role of Religion in Conflict and Peace-Building* (London: British Academy, 2015), pp. 2, 7.

79. https://twitter.com/MuslimCouncil/status/1558140294293495809?cxt=HHw WgoCx-fOVoJ8rAAAA.

   For Rushdie's justifiable earlier characterization of the threat against him (in relation to his 1988 novel *The Satanic Verses*) being a terroristic one, and for his gratitude to UK counter-terrorists for their protection, see S. Rushdie, *Joseph Anton: A Memoir* (London: Jonathan Cape, 2012), pp. 344, 353, 355, 549–50, 636.

80. J. Sacks, *Not in God's Name: Confronting Religious Violence* (London: Hodder and Stoughton, 2015).

81. D. Latimer, *A Leap of Faith: How Martin McGuinness and I Worked Together for Peace* (Newtownards: Blackstaff Press, 2018).

82. Even a less conspicuous group such as those associated with Methodism continues to comprise large numbers, there being approximately 80 million Methodists in the world (W. J. Abraham, *Methodism: A Very Short Introduction* (Oxford: OUP, 2019), pp. 1, 107).

83. M. Juergensmeyer, 'Does Religion Cause Terrorism?', in J. R. Lewis (ed.), *The Cambridge Companion to Religion and Terrorism* (Cambridge: CUP, 2017), p. 16.

84. Macdonald [Pierce], *The Turner Diaries*, pp. 35, 71–2.

85. Galatians 3:28, 5:14.

86. H. S. Gregg, *Religious Terrorism* (Cambridge: CUP, 2020).

87. J. K. Gordon, *Divine Scripture in Human Understanding: A Systematic Theology of the Christian Bible* (Notre Dame: University of Notre Dame Press, 2019), p. 6.

88. Muro (ed.), *When Does Terrorism Work?*; English, *Does Terrorism Work?*; Cronin, *How Terrorism Ends*; Abrahms, *Rules for Rebels*; M. Abrahms, 'Why Terrorism Does Not Work', *International Security*, 31/2 (2006).

89. E. Chenoweth, *Civil Resistance: What Everyone Needs to Know* (Oxford: OUP, 2021); E. Chenoweth and M. J. Stephan, *Why Civil Resistance Works: The Strategic Logic of Non-Violent Conflict* (New York: Columbia University Press, 2011).

90. Chenoweth, *Civil Resistance*, p. 127.

91. Cassam, *Extremism*, pp. 72, 152.

92. Y. Dror, 'Terrorism as a Challenge to the Democratic Capacity to Govern', in M. Crenshaw (ed.), *Terrorism, Legitimacy, and Power: The Consequences of Political Violence* (Middletown: Wesleyan University Press, 1983), p. 65.

93. C. Fabre, *Spying through a Glass Darkly: The Ethics of Espionage and Counter-Intelligence* (Oxford: OUP, 2022), p. 2.

94. Omand and Phythian, *Principled Spying.*

95. Fabre, *Spying through a Glass Darkly.*

96. Fabre, *Spying through a Glass Darkly,* pp. 199–201.

97. C. Gearty, *Terror* (London: Faber and Faber, 1991), p. 2.

98. Hassner, *Anatomy of Torture*; Soufan and Freedman, *The Black Banners (Declassified).*

99. English, *Does Terrorism Work?,* pp. 42–91.

100. See, for example, C. L. Adida, D. D. Laitin, and M. Valfort, *Why Muslim Integration Fails in Christian-Heritage Societies* (Cambridge, MA: Harvard University Press, 2016).

101. J. Argomaniz and O. Lynch (eds), *Victims of Terrorism: A Comparative and Interdisciplinary Study* (London: Routledge, 2015).

102. L. Kennedy, *Who Was Responsible for the Troubles? The Northern Ireland Conflict* (Montreal: McGill-Queen's University Press, 2020).

103. Such a view echoes the arguments of Fabre, in favour of a cosmopolitanism which holds that 'individuals each matter, and in some important sense matter equally, irrespective of their membership in this or that political community' (C. Fabre, *Cosmopolitan War* (Oxford: OUP, 2014; 1st edn 2012), p. 2).

104. Taylor, *The Ethics of Counter-Terrorism.*

105. For a summary of the principles usually involved in Just War thinking (*jus ad bellum, jus in bello, jus post bellum*), see Taylor, *The Ethics of Counter-Terrorism,* pp. 68–9.

106. For valuable reflections on the relationship between just war thinking and intelligence work, see Omand and Phythian, *Principled Spying.*

107. Fabre, *Cosmopolitan War,* p. 276.

108. And value judgement forms a legitimate part of what historians offer when they assess painful subjects such as the one under scrutiny in this book (D. Bloxham, *History and Morality* (Oxford: OUP, 2020)).

109. For subtle and historically sensitive reflection on this important theme, see Wilson, *Killing Strangers.*

110. English and Townshend (eds), *The State.*

111. Townshend, *Making the Peace.*

112. B. Blumenau, *The United Nations and Terrorism: Germany, Multilateralism, and Anti-Terrorism Efforts in the 1970s* (Basingstoke: Palgrave Macmillan, 2014).

113. L. Jarvis and T. Legrand, *Banning Them, Securing Us? Terrorism, Parliament, and the Ritual of Proscription* (Manchester: Manchester University Press, 2020).

114. D'Amato, *Cultures of Counter-Terrorism*; F. Foley, *Countering Terrorism in Britain and France: Institutions, Norms, and the Shadow of the Past* (Cambridge: CUP, 2013).

115. English, *Ernie O'Malley*, pp. 22–6, 81–2; S. McConville, *Irish Political Prisoners 1848–1922: Theatres of War* (London: Routledge, 2003); S. McConville, *Irish Political Prisoners, 1920–1962: Pilgrimage of Desolation* (London: Routledge, 2014).
116. Andrew, *The Defence of the Realm*, pp. 831, 841–2, 848–9.
117. Jensen, *The Battle against Anarchist Terrorism*.
118. Most notably D. C. Rapoport, *Waves of Global Terrorism: From 1879 to the Present* (New York: Columbia University Press, 2022).
119. Waldron, *Torture, Terror, and Trade-Offs*, p. 189.
120. English, 'Change and Continuity across the 9/11 Fault Line'.
121. See, for example, Crenshaw, *Explaining Terrorism*, pp. 53–4; Neumann, *Radicalized*, pp. xv–xvi, 173.
122. G. Orwell, *Nineteen Eighty-Four* (London: Penguin, 1954 edn), pp. 34–5.
123. English, *Terrorism: How to Respond*, pp. 52–5.
124. R. J. Evans, *In Defence of History* (London: Granta, 1997), pp. 59–61.
125. Omand, *How Spies Think*, pp. 79–84.
126. Arquilla, *Bitskrieg*, p. 24.
127. Arquilla, *Bitskrieg*, pp. 46–53.
128. Kennedy, *Who Was Responsible for the Troubles?*, pp. 69–70; English (ed.), *The Cambridge History of Terrorism*, pp. 14–15, 659, 664; English, *Armed Struggle*, pp. 145–7.
129. Kay and King, *Radical Uncertainty*.
130. English, *Does Terrorism Work?*, p. 42.
131. Powell, *Talking to Terrorists*, pp. 366–7.

Bibliographical Essay

Some readers will be more interested than others in the previous literature on counter-terrorism. This short Bibliographical Essay outlines important work that has been done by others to date and explains why this current book offers something that is original and complementary.

Does Counter-Terrorism Work? directly addresses the need for more systematic definition than currently exists about what 'working' would mean in this context, and it also aims to move beyond analyses that are over-focused on metrics. Some impressive scholars do on occasions adhere to the view that metrics are what will be required when we consider the effectiveness of counter-terrorism.[1] But, as two of the most thoughtful experts in the entire field have pointed out, 'Developing metrics for success has proved problematic', and 'Conceptions of successful counter-terrorism have varied considerably in the years since the 9/11 attacks'.[2] Other scholars have persuasively echoed such insights: 'The success of the Global War on Terror cannot simply be measured by the internment or killing of insurgents and/or terrorists'; 'the effectiveness of a counter-terrorism strategy should not be measured purely on quantitative methodologies'.[3] If the absence of a satisfactory definition of counter-terrorist efficacy therefore represents one problem with the existing literature, then the limitation of metrics-based approaches offers another.

It is also true that the polarized and sometimes tendentious nature of existing analysis is a further problem within the existing debate. Some brilliant scholars have been relentlessly critical of counter-terrorist practice and of what it has achieved. So Joseba Zulaika has proposed that, 'Counter-terrorism has become self-fulfilling and it is now pivotal in *promoting* terrorism'; 'counter-terrorism has become terrorism's best ally'.[4] Other scholars have offered analyses which resonate with this: 'terrorism has become a self-fulfilling prophecy sustained through counter-terrorist violence'; 'counter-terrorism violence helps create and perpetuate the cycles of violence';[5] 'the

---

1. Price, *Targeting Top Terrorists*, pp. 16–17.
2. Crenshaw and LaFree, *Countering Terrorism*, p. 3.
3. Tembo, *US-UK Counter-Terrorism after 9/11*, pp. 1–2.
4. Zulaika, *Terrorism: The Self-Fulfilling Prophecy*, pp. 1, 13.
5. Lindahl, *A Critical Theory of Counter-Terrorism*, pp. 1, 171. There is also a body of work which sees those who present themselves as counter-terrorist (the USA, for example) as actually being conspicuous terrorists. Pre-eminent here is Noam Chomsky (N. Chomsky, *The Culture of Terrorism* (London: Pluto Press, 1989)).

War on Terror has been an unmitigated disaster'.[6] John Mueller and Mark G. Stewart have offered coruscating critiques of counter-terrorist excess, presenting post-9/11 US counter-terrorism as having been profoundly ill-judged, excessive, wasteful, and ineffective.[7] In tune with this, others have been highly critical of what they consider to be the incompetent, financially irresponsible, and even corrupt nature of much counter-terrorist endeavour.[8] Despite the many merits of these impressive works, none of them details attentively the many *successes* that there have also been in state counter-terrorism; and this is true also of other impressive work detailing the failings in counter-terrorist effort (such as Amy Zegart's superb dissection of the organizational problems inherent within so much of the USA's security and intelligence community).[9] A full answer to the question *Does Counter-Terrorism Work?* must include the more positive aspects of state work as well.

In strong contrast to these negative assessments, some analysts have been almost bullish in their positive claims about counter-terrorist success. Thus William Matchett argues about modern-day Northern Ireland that, 'Winning the intelligence war beat the IRA'.[10] Journalist and historian Ruth Dudley Edwards says of the same conflict that the Provisional IRA experienced 'defeat', and that 'because the British and Irish security forces fought heroically and successfully to save Ireland from civil war, particularly through RUC and Garda Special Branch informers and agents within paramilitaries, ultimately the IRA had to sue for peace'.[11] Similarly, James Dingley has suggested that 'once the state kept its nerve and gave itself time to develop effective countermeasures, it was able to contain and then effectively defeat PIRA'.[12]

Of the post-9/11 US War on Terror, Peter Henne has asserted 'the general effectiveness of the US global war on terrorism'; 'the policies the United States called for as part of its global war on terrorism were aggressive and controversial, but they generally worked'.[13] Similarly, Mark Cochrane and Gabrielle Nugent have argued that the immediate post-9/11 counter-terrorist strategy of Operation Enduring Freedom 'appears successful', and that with subsequent efforts 'there has been an increasing of the national security of the United States, therefore amounting to success in terms of counter-terrorism strategies'.[14] Other analysts of the War

---

6. C. Wight, *Rethinking Terrorism: Terrorism, Violence, and the State* (London: Palgrave Macmillan, 2015), p. 219.

7. Mueller and Stewart, *Terror, Security, and Money*; Mueller and Stewart, *Chasing Ghosts*.

8. Risen, *Pay Any Price*.

9. Zegart, *Flawed by Design*; Zegart, *Spying Blind*.

10. Matchett, *Secret Victory*, p. 8.

11. *News Letter*, 8 December 2020.

12. J. Dingley, *The IRA: The Irish Republican Army* (Santa Barbara: Praeger, 2012), p. 206.

13. P. S. Henne, 'Assessing the Impact of the Global War on Terrorism on Terrorism Threats in Muslim Countries', *Terrorism and Political Violence*, 33/7 (2021), pp. 1514, 1524.

14. M. Cochrane and G. Nugent, 'Have Global Efforts to Reduce Terrorism and Political Violence since 9/11 Been Effective?', in R. Jackson and D. Pisoiu (eds), *Contemporary Debates on Terrorism* (London: Routledge, 2018; 1st edn 2012), pp. 241–2.

on Terror have sometimes exuded similar confidence. So Barry Scott Zellen, for example, in his fascinating discussion of technological innovation and post-9/11 counter-terrorism, describes 'an ability to thwart, and even defeat, terrorists, thanks to the ubiquitous nature of this new technology'; post-9/11 technology 'would provide us with a new set of tools with which to wage and, ultimately, win the continuing struggle' against terrorism.[15]

But these extremely positive evaluations might perhaps again simplify what are in fact very complex outcomes, and they surely exaggerate counter-terrorist success. For example, Henne's claims of efficacy during the War on Terror are based on a rather short-term analysis (focusing only on the years immediately after 9/11), on a rather mechanistic assessment (overly reliant on metrics), and on interpreting counter-terrorist success in terms of a rigidly narrow definition (the number of terrorist-generated deaths in each of the states covered in the article).[16]

There have also, of course, been more cautiously ambivalent assessments of the efficacy of counter-terrorism, including Daniel Byman's powerful study of Israeli experience,[17] Bruce Hoffman's extremely insightful reflections on twenty years of post-9/11 counter-terrorism,[18] or the (largely critical) analysis offered by Richard Jackson and his colleagues.[19] So too there is valuable nuance in other relevant studies, each of which yet differs significantly from this current book. Edgar Tembo offers a welcome antidote to metrics-based analysis of counter-terrorism; but his book does not provide a sustained or systematic answer to the question of the efficacy of US or UK post-9/11 counter-terrorism, nor to the question of those states' self-evaluation in terms of their own effectiveness.[20] Again, a useful Report for the US Congress in 2007 rightly observed that, 'A common pitfall of governments seeking to demonstrate success in anti-terrorist measures is over-reliance on quantitative indicators, particularly those which may correlate with progress but not accurately measure it, such as the amount of money spent on anti-terror efforts'. This Report itself, however, offered only brief and very general reflections on the difficulties of indeed assessing counter-terrorist efficacy, and it was therefore able to do little to remedy the situation.[21] Martha Crenshaw and Gary LaFree have written very impressively about how best to counter terrorism; but (understandably) their emphasis is overwhelmingly on the US rather than on wider, cross-case experience, and their focus is not sustainedly that of evaluating the efficacy of counter-terrorism

15. Zellen, *State of Recovery*, pp. 3–4.
16. Henne, 'Assessing the Impact of the Global War on Terrorism'.
17. Byman, *A High Price*.
18. B. Hoffman, 'The War on Terror 20 Years On: Crossroads or Cul-De-Sac?', Tony Blair Institute for Global Change (Commentary, 18 March 2021), https://institute.global/policy/-war-terror-20-years-crossroads-or-cul-de-sac.
19. R. Jackson, L. Jarvis, J. Gunning, and M. Breen Smyth, *Terrorism: A Critical Introduction* (Basingstoke: Palgrave Macmillan, 2011).
20. Tembo, *US-UK Counter-Terrorism after 9/11*.
21. R. Perl, *Combating Terrorism: The Challenge of Measuring Effectiveness* (Washington, DC: Congressional Research Service Report for Congress, 2007), p. 3.

historically as well as contemporaneously.[22] M. L. R. Smith and D. M. Jones have written about the (in)effectiveness of counter-insurgency, but their attention is indeed very much on counter-insurgency rather than counter-terrorism, and they do not seek to establish the kind of systematic framework for assessment that this current book provides.[23] One very valuable study addressing both counter-terrorism and counter-insurgency was offered by Hoffman and Taw in the early 1990s. Conducted for the US Department of State, their research was more emphatically intended to be practitioner-oriented than is the current book, and they aimed to outline 'the essential prerequisites to the development of national counter-terrorist/counter-insurgency strategic plans'. These authors' exhortatory framework for successful counter-terrorism therefore differs in nature from the analytical, layered framework set out in *Does Counter-Terrorism Work?*, although Hoffman and Taw's shrewd attention to the importance of coordination and cooperation within and between states seems persuasive and enduringly powerful.[24]

Some of the other best relevant analyses to date have offered concise rather than extended reflections,[25] and so again they differ from this current study. One short essay by Sinai, Fuller, and Seal usefully reviewed the literature on measures of effectiveness in countering violent extremism and terrorism. But their brief over-view necessarily tended towards very general points indeed about efficacy: 'to be effective, CT [counter-terrorism] interventions need to understand the nature of the terrorist threats in terms of their physical geographical location as domestic, transnational, international, state, or in cyberspace. It is also important to understand the types of adversary groups, whether these are hierarchically organized, loosely networked, lone-actors, or state-sponsored, or a combination of these types. Counter-terrorism campaigns also need to understand the nature of adversaries as political, religious, criminal, or other types, as well as their linkages with other terrorist groups.' The review did, however, helpfully draw attention to recognition within the existing literature that there exists no agreement about what success in counter-terrorism actually involves—a very important problem, and one which this current book centrally addresses.[26] Enders and Sandler sharp-sightedly assessed a small number of tactical measures used in then-recent US counter-terrorism (metal detectors in airports; embassy fortification; the increasing of severity of punishment for terrorist acts through two new laws introduced in 1984; a single retaliatory

22. Crenshaw and LaFree, *Countering Terrorism*.
23. M. L. R. Smith and D. M. Jones, *The Political Impossibility of Modern Counterinsurgency: Strategic Problems, Puzzles, and Paradoxes* (New York: Columbia University Press, 2015).
24. B. Hoffman and J. M. Taw, *A Strategic Framework for Countering Terrorism and Insurgency* (Santa Monica: RAND, 1992), quotation at p. iv.
25. B. Hoffman, 'Rethinking Terrorism and Counter-Terrorism since 9/11', *Studies in Conflict and Terrorism*, 25/5 (2002); A. Roberts, 'The "War on Terror" in Historical Perspective', *Survival*, 47/2 (2005).
26. J. Sinai, J. Fuller, and T. Seal, 'Research Note: Effectiveness in Counter-Terrorism and Countering Violent Extremism: A Literature Review', *Perspectives on Terrorism*, 13/6 (2019), quotation at p. 90.

attack—on Libya—in 1986); again, the range of (and frameworks for) analysis are therefore much narrower in their article than is the case in this current book.[27]

Ronald Crelinsten's valuable *Counterterrorism*[28] offers a useful overview of various approaches that can be adopted towards terrorism (coercive, proactive, persuasive, defensive, and long-term varieties of counter-terrorism), and considers some of the strengths and weaknesses of each approach within this inventory. Crelinsten's aim is therefore somewhat different from my own. He concludes by advocating a counter-terrorism which draws on the various kinds of counter-terrorism that he discusses; he does not intend to set out the kind of layered framework for analysis of counter-terrorist efficacy offered in this current book, then systematically and sustainedly to test it against detailed case studies viewed historically. Indeed, unlike *Does Counter-Terrorism Work?*, Crelinsten's impressive book somewhat leaves aside the issue of actual efficacy, of how the various modes and methods of counter-terrorism that he lists actually work in contextual and detailed practice.[29]

Somewhat similarly, Christopher Hewitt's valuable, short 1984 study *The Effectiveness of Anti-Terrorist Policies* concisely reflects on five cases of urban terrorism, assessing in each instance the effectiveness of a series of anti-terrorism policies (negotiating a ceasefire; improving the economic situation; introducing reforms; using collective punishment; deploying emergency legislation; utilizing security force and other repressive measures). The timelines are much more short-term than those considered in this current book, Hewitt analysing Northern Ireland 1970–81, Spain 1975–81, Italy 1977–81, Uruguay 1968–73, and Cyprus 1955–8. This clearly limits the extent to which he can evaluate the longer-term effects of counter-terrorist approaches (the combination, for example, of reform, negotiation, and security force activity), a limitation reinforced by the fact that much of his analysis considers either immediate effects or very short ones over a matter of only a few months. All of this (together with Hewitt's deciding not to offer a layered framework for what counter-terrorist effectiveness might mean in practice) makes his book significantly different from *Does Counter-Terrorism Work?*, in approach but also in conclusions.[30]

Other valuable publications examining counter-terrorism again involve people adopting relevant but somewhat different agendas. Robert Art and Louise Richardson co-edited a marvellous 2007 volume (*Democracy and Counter-Terrorism*), in which they and their various authors asked how different democratic states had combatted terrorism in recent decades, and what lessons might be drawn from these

---

27. W. Enders and T. Sandler, 'The Effectiveness of Anti-Terrorism Policies: A Vector-Autoregression-Intervention Analysis', *American Political Science Review*, 87/4 (1993).
28. Crelinsten, *Counter-Terrorism*.
29. See, for example, Crelinsten, *Counter-Terrorism*, pp. 58, 237, 239, 241, 243–6; 'The question remains as to how all this works in reality' (p. 242).
30. C. Hewitt, *The Effectiveness of Anti-Terrorist Policies* (Lanham: University Press of America, 1984). Hewitt's suggestion, for example, that, 'Negotiations will not lead to resolution of the conflict' (p. 41) is not one that I share.

experiences for counter-terrorism in the present and the future. It remains an important study, its different chapter authors each reflecting on an individual country case. Reasonably enough, given the editors' priorities, the book did not set out to establish a systematic framework for evaluating what counter-terrorist success would involve in detail, a framework on which basis the authors would synoptically interrogate each case study. This is not a flaw in the book; but it does mean that the aim and scope of that impressive volume are different from those embodied in *Does Counter-Terrorism Work?*[31]

Yet again, an insightful book such as Beatrice de Graaf's comparative assessment of the 1970s focuses only on one short period, and also concentrates not on the overall efficacy of counter-terrorism, but rather on something else. In Professor de Graaf's view, it is 'the theatrical quality of both terrorism *and the ensuing measures* that define their social impact', and her interest is in the 'performative power' of counter-terrorism:

> Thus, the performative power of counter-terrorism can be defined as the extent to which the national government, by means of its official counter-terrorism policy and corresponding discourse (in statements, enactments, measures, and ministers' remarks) aims to mobilize public and political support and in the last instance, wittingly or unwittingly, assists the terrorists in creating social drama.

With this focus, she distances herself from 'the technical questions about counter-terrorism effectiveness that are epistemologically or empirically almost impossible to answer'; given this remit, de Graaf understandably avoids systematic definition of what effective counter-terrorism would involve, thereby taking a very different approach from that adopted in the current book. I agree that counter-terrorist 'effectiveness is hard to assess'.[32] But the argument of *Does Counter-Terrorism Work?* is that it is indeed possible to assess it meaningfully and rigorously. What is needed is a balanced rather than tendentious approach, and one based on a systematic framework of analysis: on a layered definition of what effective counter-terrorism would involve. Further, such an approach involves testing that framework through cross-case, historically informed comparison of lived experience in a book-length treatment. None of the valuable works referred to in this Bibliographical Essay does this,[33] but I hope that the current book has made some contribution towards this important goal.

31. R. J. Art and L. Richardson (eds), *Democracy and Counter-Terrorism: Lessons from the Past* (Washington, DC: United States Institute of Peace Press, 2007).
32. De Graaf, *Evaluating Counter-Terrorism Performance*, pp. 7, 10–12.
33. And nor is it the intention of other superb books in the field to do so, including Crenshaw, *Explaining Terrorism*; Crenshaw (ed.), *The Consequences of Counter-Terrorism*; and Ganor, *The Counter-Terrorism Puzzle*.

Bibliography

Abraham, W. J., *Methodism: A Very Short Introduction* (Oxford: OUP, 2019).

Abrahms, M., 'Al-Qaida's Scorecard: A Progress Report on Al-Qaida's Objectives', *Studies in Conflict and Terrorism*, 29/5 (2006).

Abrahms, M., 'Why Terrorism Does Not Work', *International Security*, 31/2 (2006).

Abrahms, M., *Rules for Rebels: The Science of Victory in Militant History* (Oxford: OUP, 2018).

Abu-Bakare, A., 'Counter-Terrorism and Race', *International Politics Reviews*, 8 (2020).

Ackerman, G. A., 'The Provisional Irish Republican Army and the Development of Mortars', *Journal of Strategic Security*, 9/1 (2016).

Adams, G., *The Politics of Irish Freedom* (Dingle: Brandon, 1986).

Adams, G., *A Pathway to Peace* (Cork: Mercier Press, 1988).

Adida, C. L., Laitin, D. D., and Valfort, M., *Why Muslim Integration Fails in Christian-Heritage Societies* (Cambridge, MA: Harvard University Press, 2016).

Alonso, R., *The IRA and Armed Struggle* (London: Routledge, 2007).

Alonso, R., 'Why Do Terrorists Stop? Analyzing Why ETA Members Abandon or Continue with Terrorism', *Studies in Conflict and Terrorism*, 34/9 (2011).

Andrew, C., *The Defence of the Realm: The Authorized History of MI5* (London: Penguin, 2009).

Angstrom, J., and Duyvesteyn, I. (eds), *Understanding Victory and Defeat in Contemporary War* (London: Routledge, 2007).

Argomaniz, J., and Lynch, O. (eds), *Victims of Terrorism: A Comparative and Interdisciplinary Study* (London: Routledge, 2015).

Arquilla, J., *Bitskrieg: The New Challenge of Cyberwarfare* (Cambridge: Polity Press, 2021).

Arsenault, E. G., *How the Gloves Came Off: Lawyers, Policy-Makers, and Norms in the Debate on Torture* (New York: Columbia University Press, 2017).

Art, R. J., and Richardson, L. (eds), *Democracy and Counter-Terrorism: Lessons from the Past* (Washington, DC: United States Institute of Peace Press, 2007).

Ashworth, S., Berry, C. R., and Bueno de Mesquita, E., *Theory and Credibility: Integrating Theoretical and Empirical Social Science* (Princeton: Princeton University Press, 2021).

Augusteijn, J. (ed.), *The Irish Revolution, 1913–1923* (Basingstoke: Palgrave Macmillan, 2002).

Baddiel, D., *Jews Don't Count: How Identity Politics Failed One Particular Identity* (London: TLS Books, 2021).

Bamford, B. W. C., 'The Role and Effectiveness of Intelligence in Northern Ireland', *Intelligence and National Security*, 20/4 (2005).

Barfield, T., *Afghanistan: A Cultural and Political History* (Princeton: Princeton University Press, 2012; 1st edn 2010).

Begin, M., *The Revolt: Story of the Irgun* (Jerusalem: Steimatzky, 1977; 1st edn 1952).

Bell, E., Owen, T., Khorana, S., and Henrichsen, J. R. (eds), *Journalism after Snowden: The Future of the Free Press in the Surveillance State* (New York: Columbia University Press, 2017).

Bentley, M., *Modern Historiography: An Introduction* (London: Routledge, 1999).

Bergen, P., *Manhunt: From 9/11 to Abbottabad—The Ten-Year Search for Osama bin Laden* (London: Bodley Head, 2012).

Berger, J. M., *The Turner Legacy: The Storied Origins and Enduring Impact of White Nationalism's Deadly Bible* (The Hague: ICCT, 2016).

Bergia, E., 'Unexpected Rewards of Political Violence: Republican Ex-Prisoners, Seductive Capital, and the Gendered Nature of Heroism', *Terrorism and Political Violence*, 33/7 (2021).

Berman, E., *Radical, Religious, and Violent: The New Economics of Terrorism* (Cambridge, MA: MIT Press, 2009).

Berman, E., Felter, J. H., and Shapiro, J. N., *Small Wars, Big Data: The Information Revolution in Modern Conflict* (Princeton: Princeton University Press, 2018).

Berrebi, C., and Klor, E. F., 'On Terrorism and Electoral Outcomes: Theory and Evidence from the Israeli-Palestinian Conflict', *Journal of Conflict Resolution*, 50/6 (2006).

Berrebi, C., and Klor, E. F., 'Are Voters Sensitive to Terrorism? Direct Evidence from the Israeli Electorate', *American Political Science Review*, 102/3 (2008).

Berrebi, C., and Klor, E. F., 'The Impact of Terrorism on the Defence Industry', *Economica*, 77/307 (2010).

Bew, J., Frampton, M., and Gurruchaga, I., *Talking to Terrorists: Making Peace in Northern Ireland and the Basque Country* (London: Hurst and Company, 2009).

Bew, J., and Frampton, M., '"Don't mention the war!": Debating Notion of a "Stalemate" in Northern Ireland (and a Response to Dr Paul Dixon)', *Journal of Imperial and Commonwealth History*, 40/2 (2012).

Bew, P., *Ireland: The Politics of Enmity 1789–2006* (Oxford: OUP, 2007).

Biggar, N., *In Defence of War* (Oxford: OUP, 2013).

Bishop, P., *The Man Who Was Saturday: The Extraordinary Life of Airey Neave—Soldier, Escaper, Spymaster, Politician* (London: William Collins, 2019).

Bishop, P., and Mallie, E., *The Provisional IRA* (London: Corgi, 1988; 1st edn 1987).

Blair, T., *A Journey* (London: Hutchinson, 2010).

Blattman, C., *Why We Fight: The Roots of War and the Paths to Peace* (New York: Viking, 2022).

Bloxham, D., *History and Morality* (Oxford: OUP, 2020).

Blumenau, B., *The United Nations and Terrorism: Germany, Multilateralism, and Anti-Terrorism Efforts in the 1970s* (Basingstoke: Palgrave Macmillan, 2014).

Bowler, P. J., *Progress Unchained: Ideas of Evolution, Human History, and the Future* (Cambridge: CUP, 2021).

Boyce, D. G., *Nineteenth-Century Ireland: The Search for Stability* (Dublin: Gill and Macmillan, 1990).

Boyce, D. G. (ed.), *The Revolution in Ireland, 1879–1923* (Basingstoke: Macmillan, 1988).

Boyle, M. J., *The Drone Age: How Drone Technology Will Change War and Peace* (Oxford: OUP, 2020).

Bradley, G., and Feeney, B., *Insider: Gerry Bradley's Life in the IRA* (Dublin: O'Brien Press, 2009).

Brahimi, A., *Jihad and Just War in the War on Terror* (Oxford: OUP, 2010).

Brennan, M., *The War in Clare 1911–1921: Personal Memoirs of the Irish War of Independence* (Dublin: Four Courts Press, 1980).

Brewer, J. D., *Advanced Introduction to the Sociology of Peace Processes* (Cheltenham: Edward Elgar, 2022).

Brewer, J. D., and Magee, K., *Inside the RUC: Routine Policing in a Divided Society* (Oxford: OUP, 1991).

Bruce, S., *The Red Hand: Protestant Paramilitaries in Northern Ireland* (Oxford: OUP, 1992).

Burke, E., *An Army of Tribes: British Army Cohesion, Deviancy, and Murder in Northern Ireland* (Liverpool, Liverpool University Press, 2018).

Burke, J., *The 9/11 Wars* (London: Penguin, 2011).

Byman, D., *A High Price: The Triumphs and Failures of Israeli Counter-Terrorism* (Oxford: OUP, 2011).

Byman, D., 'The Good Enough Doctrine: Learning to Live with Terrorism', *Foreign Affairs*, 100/5 (2021).

Cadwallader, A., *Lethal Allies: British Collusion in Ireland* (Cork: Mercier Press, 2013).

Cameron, D., *For the Record* (London: William Collins, 2019).

Cannadine, D., *The Undivided Past: History beyond Our Differences* (London: Penguin, 2013).

Carlin, W., *Thatcher's Spy: My Life as an MI5 Agent inside Sinn Fein* (Newbridge: Merrion Press, 2021; 1st edn 2019).

Cassam, Q., *Conspiracy Theories* (Cambridge: Polity Press, 2019).

Cassam, Q., *Vices of the Mind: From the Intellectual to the Political* (Oxford: OUP, 2019).

Cassam, Q., *Extremism: A Philosophical Analysis* (London: Routledge, 2022).

Chard, D. S., *Nixon's War at Home: The FBI, Leftist Guerrillas, and the Origins of Counter-Terrorism* (Chapel Hill: University of North Carolina Press, 2021).

Chenoweth, E., *Civil Resistance: What Everyone Needs to Know* (Oxford: OUP, 2021).

Chenoweth, E., and Stephan, M. J., *Why Civil Resistance Works: The Strategic Logic of Non-Violent Conflict* (New York: Columbia University Press, 2011).

Chenoweth, E., English, R., Gofas, A., and Kalyvas, S. N. (eds), *The Oxford Handbook of Terrorism* (Oxford: OUP, 2019).

Choi, S., 'Fighting Terrorism through the Rule of Law?', *Journal of Conflict Resolution*, 54/6 (2010).

Chomsky, N., *The Culture of Terrorism* (London: Pluto Press, 1989).

Christia, F., *Alliance Formation in Civil Wars* (Cambridge: CUP, 2012).

Clark, J. C. D., *Our Shadowed Present: Modernism, Postmodernism, and History* (London: Atlantic Books, 2003).

Clarke, R. A., *Against All Enemies: Inside America's War on Terror* (London: Free Press, 2004).

Clinton, H. R., and Penny, L., *State of Terror* (London: Pan Macmillan, 2021).

Cobain, I., *Anatomy of a Killing: Life and Death on a Divided Island* (London: Granta, 2020).

Coll, S., *Directorate S: The CIA and America's Secret Wars in Afghanistan and Pakistan, 2001–2016* (London: Penguin, 2019; 1st edn 2018).

Collins, E. (with McGovern, M.), *Killing Rage* (London: Granta, 1997).

Comey, J., *A Higher Loyalty: Truth, Lies, and Leadership* (London: Macmillan, 2018).

Coogan, T. P., *The IRA* (London: Fontana, 1987; 1st edn 1970).

Cottam, M. L., Huseby, J. W., and Baltodano, B., *Confronting al-Qaida: The Sunni Awakening and American Strategy in al Anbar* (Lanham: Rowman and Littlefield, 2016).

Craig, T., '"You Will be Responsible to the GOC": Stovepiping and the Problem of Divergent Intelligence Gathering Networks in Northern Ireland, 1969–1975', *Intelligence and National Security*, 33/2 (2018).

Crawford, C., *Inside the UDA: Volunteers and Violence* (London: Pluto Press, 2003).

Crawley, J., *The Yank: My Life as a Former US Marine in the IRA* (Newbridge: Merrion Press, 2022).

Crelinsten, R., *Counter-Terrorism* (Cambridge: Polity Press, 2009).

Crenshaw, M., *Explaining Terrorism: Causes, Processes, and Consequences* (London: Routledge, 2011).

Crenshaw, M. (ed.), *Terrorism, Legitimacy, and Power: The Consequences of Political Violence* (Middletown: Wesleyan University Press, 1983).

Crenshaw, M. (ed.), *Terrorism in Context* (University Park: Pennsylvania State University Press, 1995).

Crenshaw, M. (ed.), *The Consequences of Counter-Terrorism* (New York: Russell Sage Foundation, 2010).

Crenshaw, M., and LaFree, G., *Countering Terrorism* (Washington, DC: Brookings Institution Press, 2017).

Croft, S., *Securitizing Islam: Identity and the Search for Security* (Cambridge: CUP, 2012).

Cronin, A. K., *How Terrorism Ends: Understanding the Decline and Demise of Terrorist Campaigns* (Princeton: Princeton University Press, 2009).

Cunningham, M. J., *British Government Policy in Northern Ireland 1969–89: Its Nature and Execution* (Manchester: Manchester University Press, 1991).

Cunningham, M. J., *British Government Policy in Northern Ireland, 1969–2000* (Manchester: Manchester University Press, 2001).

Currie, P. M., and Taylor, M. (eds), *Dissident Irish Republicanism* (London: Continuum, 2011).

Cursey, C., *MRF Shadow Troop* (London: Thistle Publishing, 2013).

Curtin, N. J., *The United Irishmen: Popular Politics in Ulster and Dublin 1791–1798* (Oxford: OUP, 1994).

Cusack, J., and McDonald, H., *UVF* (Dublin: Poolbeg Press, 2000; 1st edn 1997).

D'Amato, S., *Cultures of Counter-Terrorism: French and Italian Responses to Terrorism After 9/11* (London: Routledge, 2019).

David, S., *Operation Thunderbolt: Flight 139 and the Raid on Entebbe Airport, the Most Audacious Hostage Rescue Mission in History* (London: Hodder and Stoughton, 2015).

Davis, J. (ed.), *Africa and the War on Terrorism* (Aldershot: Ashgate, 2007).

de Breadun, D., *The Far Side of Revenge: Making Peace in Northern Ireland* (Cork: Collins Press, 2008; 1st edn 2001).

de Graaf, B., *Evaluating Counter-Terrorism Performance: A Comparative Study* (London: Routledge, 2013; 1st edn 2011).

Dershowitz, A. M., *Why Terrorism Works: Understanding the Threat, Responding to the Challenge* (New Haven: Yale University Press, 2002).

Dickson, D., Keogh, D., and Whelan, K. (eds), *The United Irishmen: Republicanism, Radicalism, and Rebellion* (Dublin: Lilliput Press, 1993).

Dingley, J., *The IRA: The Irish Republican Army* (Santa Barbara: Praeger, 2012).

Dixon, P., 'Guns First, Talks Later: Neoconservatives and the Northern Ireland Peace Process', *Journal of Imperial and Commonwealth History*, 39/4 (2011).

Dixon, P., 'Was the IRA Defeated? Neo-Conservative Propaganda as History', *Journal of Imperial and Commonwealth History*, 40/2 (2012).

Donohue, L. K., *The Cost of Counter-Terrorism: Power, Politics, and Liberty* (Cambridge: CUP, 2008).

Dorani, S., *America in Afghanistan: Foreign Policy and Decision Making from Bush to Obama to Trump* (London: I. B. Tauris, 2019).

Drower, G., *John Hume: Peacemaker* (London: Victor Gollancz, 1995).

Edwards, A., *UVF: Behind the Mask* (Newbridge: Merrion Press, 2017).

Edwards, A., *Agents of Influence: Britain's Secret Intelligence War against the IRA* (Newbridge: Merrion Press, 2021).

Enders, W., and Sandler, T., 'The Effectiveness of Anti-Terrorism Policies: A Vector-Autoregression-Intervention Analysis', *American Political Science Review*, 87/4 (1993).

English, D., *Christianity and Politics* (Belfast: Queen's University Belfast, 1993).

English, R., *Radicals and the Republic: Socialist Republicanism in the Irish Free State 1925–1937* (Oxford: OUP, 1994).

English, R., *Ernie O'Malley: IRA Intellectual* (Oxford: OUP, 1998).

English, R., *Irish Freedom: The History of Nationalism in Ireland* (London: Pan Macmillan, 2006).

English, R., *Terrorism: How to Respond* (Oxford: OUP, 2009).

English, R., 'Directions in Historiography: History and Irish Nationalism', *Irish Historical Studies*, 37/147 (2011).

English, R., *Armed Struggle: The History of the IRA* (London: Pan Macmillan, 2012; 1st edn 2003).

English, R., *Modern War: A Very Short Introduction* (Oxford: OUP, 2013).

English, R., *Does Terrorism Work? A History* (Oxford: OUP, 2016).

English, R., 'The Future Study of Terrorism', *European Journal of International Security*, 1/2 (2016).

English, R., 'Change and Continuity across the 9/11 Fault Line: Rethinking Twenty-First-Century Responses to Terrorism', *Critical Studies on Terrorism*, 12/1 (2019).

English, R. (ed.), *Illusions of Terrorism and Counter-Terrorism* (Oxford: OUP, 2015).

English, R. (ed.), *The Cambridge History of Terrorism* (Cambridge: CUP, 2021).

English, R., and Townshend, C. (eds), *The State: Historical and Political Dimensions* (London: Routledge, 1999).

Evans, R. J., *In Defence of History* (London: Granta, 1997).

Evans, R. J., *Altered Pasts: Counterfactuals in History* (London: Little, Brown, 2014).

Fabre, C., *Cosmopolitan War* (Oxford: OUP, 2014; 1st edn 2012).

Fabre, C., *Spying through a Glass Darkly: The Ethics of Espionage and Counter-Intelligence* (Oxford: OUP, 2022).

Fair, C. C., Kaltenthaler, K., and Miller, W. J., 'Pakistani Opposition to American Drone Strikes', *Political Science Quarterly*, 129/1 (2014).

Fellman, M., *In the Name of God and Country: Reconsidering Terrorism in American History* (New Haven: Yale University Press, 2010).

Ferriter, D., *The Border: The Legacy of a Century of Anglo-Irish Politics* (London: Profile Books, 2019).

Finkelstein, N. G., *Gaza: An Inquest into Its Martyrdom* (Oakland: University of California Press, 2018).

FitzGerald, G., *All in a Life: An Autobiography* (Dublin: Gill and Macmillan, 1992; 1st edn 1991).

FitzGerald, G., 'The 1974–5 Threat of a British Withdrawal from Northern Ireland', *Irish Studies in International Affairs*, 17 (2006).

Flackes, W. D., and Elliott, S., *Northern Ireland: A Political Directory 1968–1993* (Belfast: Blackstaff Press, 1994; 1st edn 1980).

Flanagan, F., *Remembering the Revolution: Dissent, Culture, and Nationalism in the Irish Free State* (Oxford: OUP, 2015).

Foley, F., *Countering Terrorism in Britain and France: Institutions, Norms, and the Shadow of the Past* (Cambridge: CUP, 2013).

Foster, R. F., *Vivid Faces: The Revolutionary Generation in Ireland 1890–1923* (London: Penguin, 2014).

Frampton, M., *Legion of the Rearguard: Dissident Irish Republicanism* (Dublin: Irish Academic Press, 2011).

Freedman, L., *Strategy: A History* (Oxford: OUP, 2013).

Freilich, C. D., 'Israel's Counter-Terrorism Policy: How Effective?', *Terrorism and Political Violence*, 29/2 (2017).

Fukuyama, F., *After the Neocons: America at the Crossroads* (London: Profile, 2006).

Ganiel, G., 'Praying for Paisley—Fr Gerry Reynolds and the Role of Prayer in Faith-Based Peacebuilding: A Preliminary Theoretical Framework', *Irish Political Studies*, 36/1 (2021).

Ganor, B., *The Counter-Terrorism Puzzle: A Guide for Decision Makers* (London: Routledge, 2017; 1st edn 2005).

Ganor, B., *Israel's Counter-Terrorism Strategy: Origins to the Present* (New York: Columbia University Press, 2021).

Gantt, J., *Irish Terrorism in the Atlantic Community, 1865–1922* (Basingstoke: Palgrave Macmillan, 2010).

Gearty, C., *Terror* (London: Faber and Faber, 1991).

Gearty, C., *Can Human Rights Survive?* (Cambridge: CUP, 2006).

Gearty, C., *Civil Liberties* (Oxford: OUP, 2007).

Gearty, C., *Liberty and Security* (Cambridge: Polity Press, 2013).

Gelvin, J. L., *The Modern Middle East: A History* (Oxford: OUP, 2016; 1st edn 2005).

Gerges, F. A., *ISIS: A History* (Princeton: Princeton University Press, 2016).

Gill, P., *Lone-Actor Terrorists: A Behavioural Analysis* (London: Routledge, 2015).

Gillespie, G., *Years of Darkness: The Troubles Remembered* (Dublin: Gill and Macmillan, 2008).

Godson, D., *Himself Alone: David Trimble and the Ordeal of Unionism* (London: Harper Perennial, 2005; 1st edn 2004).

Gordon, J. K., *Divine Scripture in Human Understanding: A Systematic Theology of the Christian Bible* (Notre Dame: University of Notre Dame Press, 2019).

Gorka, S., *Defeating Jihad: The Winnable War* (Washington, DC: Regnery Publishing, 2016).

Grady, J., *Six Days of the Condor* (Harpenden: No Exit Press, 2015; 1st edn 1974).

Gregg, H. S., *Religious Terrorism* (Cambridge: CUP, 2020).

Guelke, A., and Wright, F., 'The Option of a "British Withdrawal" from Northern Ireland: An Exploration of Its Meaning, Influence, and Feasibility', *Conflict Quarterly*, 10/4 (1990).

Guldi, J., and Armitage, D., *The History Manifesto* (Cambridge: CUP, 2014).

Gunning, J., *Hamas in Politics: Democracy, Religion, Violence* (London: Hurst and Co., 2009).

Haass, R. N., *War of Necessity, War of Choice: A Memoir of Two Iraq Wars* (New York: Simon and Schuster, 2010; 1st edn 2009).

Hadad, O., 'A Battle of Names: Hamas and Israeli Operations in the Gaza Strip', *Terrorism and Political Violence*, 33/5 (2021).

Hafez, M. M., and Hatfield, J. M., 'Do Targeted Assassinations Work? A Multivariate Analysis of Israel's Controversial Tactic during Al-Aqsa Uprising', *Studies in Conflict and Terrorism*, 29/4 (2006).

Hamilton, B., *Making Sense of the Bible: Rediscovering the Power of Scripture Today* (New York: HarperCollins, 2014).

Harris-Gershon, D., *What Do You Buy the Children of the Terrorist Who Tried to Kill Your Wife? A Memoir* (London: Oneworld Publications, 2013).

Hassan, R., *Life as a Weapon: The Global Rise of Suicide Bombings* (London: Routledge, 2011).

Hassner, R. E., *Anatomy of Torture* (Ithaca: Cornell University Press, 2022).

Hayes, P. (ed.), *Themes in Modern European History* (London: Routledge, 1992).

Henne, P. S., 'Assessing the Impact of the Global War on Terrorism on Terrorism Threats in Muslim Countries', *Terrorism and Political Violence*, 33/7 (2021).

Hepworth, J., *'The Age-Old Struggle': Irish Republicanism from the Battle of the Bogside to the Belfast Agreement, 1969–1998* (Liverpool: Liverpool University Press, 2021).

Hersh, S. M. *Chain of Command: The Road from 9/11 to Abu Ghraib* (New York: HarperCollins, 2004).

Hewitt, C., *The Effectiveness of Anti-Terrorist Policies* (Lanham: University Press of America, 1984).

Hewitt, S., *The British War on Terror: Terrorism and Counter-Terrorism on the Home Front since 9/11* (London: Continuum, 2008).

Hewitt, S., *Snitch! A History of the Modern Intelligence Informer* (London: Continuum, 2010).

Hobbes, T., *Leviathan* (London: Fontana, 1962; 1st edn 1651).

Hobsbawm, E., *On History* (London: Weidenfeld and Nicolson, 1997).

Hoffman, B., 'Rethinking Terrorism and Counter-Terrorism since 9/11', *Studies in Conflict and Terrorism*, 25/5 (2002).

Hoffman, B., *Anonymous Soldiers: The Struggle for Israel, 1917–1947* (New York: Alfred A. Knopf, 2015).

Hoffman, B., *Inside Terrorism* (New York: Columbia University Press, 2017; 1st edn 1998).

Hoffman, B., and Taw, J. M., *A Strategic Framework for Countering Terrorism and Insurgency* (Santa Monica: RAND, 1992).

Hoffman, B., and Reinares, F. (eds), *The Evolution of the Global Terrorist Threat: From 9/11 to Osama bin Laden's Death* (New York: Columbia University Press, 2014).

Holland, J., and Phoenix, S., *Phoenix: Policing the Shadows. The Secret War against Terrorism in Northern Ireland* (London: Hodder and Stoughton, 1997; 1st edn 1996).

Horgan, J., *The Psychology of Terrorism* (London: Routledge, 2005).

Horgan, J., *Walking Away from Terrorism: Accounts of Disengagement from Radical and Extremist Movements* (London: Routledge, 2009).

Horgan, J., *Divided We Stand: The Strategy and Psychology of Ireland's Dissident Terrorists* (Oxford: OUP, 2013).

Ingram, M., and Harkin, G., *Stakeknife: Britain's Secret Agents in Ireland* (Dublin: O'Brien Press, 2004).

Jackson, A., *Ireland 1798–1998: War, Peace, and Beyond* (Chichester: Wiley-Blackwell, 2010; 1st edn 1999).

Jackson, R., Breen Smyth, M., and Gunning, J. (eds), *Critical Terrorism Studies: A New Research Agenda* (London: Routledge, 2009).

Jackson, R., Jarvis, L., Gunning, J., and Breen Smyth, M., *Terrorism: A Critical Introduction* (Basingstoke: Palgrave Macmillan, 2011).

Jackson, R., and Pisoiu, D. (eds), *Contemporary Debates on Terrorism* (London: Routledge, 2018; 1st edn 2012).

Jarvis, L., and Legrand, T., *Banning Them, Securing Us? Terrorism, Parliament, and the Ritual of Proscription* (Manchester: Manchester University Press, 2020).

Jensen, R. B., *The Battle against Anarchist Terrorism: An International History, 1878–1934* (Cambridge: CUP, 2014).

Johnson, T. H., *Taliban Narratives: The Use and Power of Stories in the Afghanistan Conflict* (London: Hurst and Co., 2017).

Jones, L. G., *Christian Social Innovation: Renewing Wesleyan Witness* (Nashville: Abingdon Press, 2016).

Jones, S. G., *Counter-Insurgency in Afghanistan* (Santa Monica: RAND, 2008).

Kaplan, E. H., Mintz, A., Mishal, S., and Samban, C., 'What Happened to Suicide Bombings in Israel? Insights from a Terror Stock Model', *Studies in Conflict and Terrorism*, 28/3 (2005).

Kay, J., and King, M., *Radical Uncertainty: Decision-Making for an Unknowable Future* (London: Bridge Street Press, 2020).

Kelly, M. J., *The Fenian Ideal and Irish Nationalism, 1882–1916* (Woodbridge: Boydell Press, 2006).

Kelly, S., *Margaret Thatcher, the Conservative Party, and the Northern Ireland Conflict, 1975–1990* (London: Bloomsbury, 2021).

Kennedy, L., *Who Was Responsible for the Troubles? The Northern Ireland Conflict* (Montreal: McGill-Queen's University Press, 2020).

Kerr, M., *The Destructors: The Story of Northern Ireland's Lost Peace Process* (Dublin: Irish Academic Press, 2011).

Khalidi, R., *Palestinian Identity: The Construction of Modern National Consciousness* (New York: Columbia University Press, 2010; 1st edn 1997).

Kilcullen, D., *Blood Year: Islamic State and the Failures of the War on Terror* (London: Hurst and Co., 2016).

Kirk-Smith, M., and Dingley, J., 'Countering Terrorism in Northern Ireland: The Role of Intelligence', *Small Wars and Insurgencies*, 20/3–4 (2009).

Kitson, F., *Low Intensity Operations: Subversion, Insurgency, and Peacekeeping* (London: Faber and Faber, 2010; 1st edn 1971).

Klein, K. L., *From History to Theory* (Berkeley: University of California Press, 2012).

Knatchbull, T., *From a Clear Blue Sky: Surviving the Mountbatten Bomb* (London: Hutchinson, 2009).

Knox, D. B., *The Killing of Thomas Niedermayer* (Stillorgan: New Island Books, 2019).

Krause, P., *Rebel Power: Why National Movements Compete, Fight, and Win* (Ithaca: Cornell University Press, 2017).

Kruglova, A., *Terrorist Recruitment, Propaganda, and Branding: Selling Terror Online* (London: Routledge, 2023).

Kurtz-Phelan, D., 'Who Won the War on Terror?', *Foreign Affairs*, 100/5 (2021).

Kurtzer, D. C., Lasensky, S. B., Quandt, W. B., Spiegel, S. L., and Telhami, S. Z., *The Peace Puzzle: America's Quest for Arab-Israeli Peace, 1989–2011* (Ithaca: Cornell University Press, 2013).

LaFree, G., Dugan, L., and Miller, E., *Putting Terrorism in Context: Lessons from the Global Terrorism Database* (London: Routledge, 2015).

Lahoud, N., 'Bin Laden's Catastrophic Success: Al-Qaida Changed the World—But Not in the Way It Expected', *Foreign Affairs*, 100/5 (2021).

Latimer, D., *A Leap of Faith: How Martin McGuinness and I Worked Together for Peace* (Newtownards: Blackstaff Press, 2018).

Lawrence, B. (ed.), *Messages to the World: The Statements of Osama bin Laden* (London: Verso, 2005).

Leahy, T., *The Intelligence War against the IRA* (Cambridge: CUP, 2020).

Lenard, P. T., *How Should Democracies Fight Terrorism?* (Cambridge: Polity Press, 2020).

Levi, M., *On Nuclear Terrorism* (Cambridge, MA: Harvard University Press, 2007).

Levitt, M., *Hamas: Politics, Charity, and Terrorism in the Service of Jihad* (New Haven: Yale University Press, 2006).

Lewis, J. R. (ed.), *The Cambridge Companion to Religion and Terrorism* (Cambridge: CUP, 2017).

Lieven, A., *Pakistan: A Hard Country* (London: Penguin, 2012; 1st edn 2011).

Lindahl, S., *A Critical Theory of Counter-Terrorism: Ontology, Epistemology, and Normativity* (London: Routledge, 2018).

Lodge, J. (ed.), *Terrorism: A Challenge to the State* (Oxford: Martin Robertson, 1981).

Lowe, D., *Policing Terrorism: Research Studies into Police Counter-Terrorism Investigations* (London: CRC Press, 2019; 1st edn 2016).

Luft, G., 'The Logic of Israel's Targeted Killing', *Middle East Quarterly*, 10/1 (2003).

Lum, C., Kennedy, L. W., and Sherley, A., 'Are Counter-Terrorism Strategies Effective? The Results of the Campbell Systematic Review on Counter-Terrorism Evaluation Research', *Journal of Experimental Criminology*, 2/4 (2006).

Maillot, A., *Rebels in Government: Is Sinn Fein Ready for Power?* (Manchester: Manchester University Press, 2022).

Malcolm, N., *Aspects of Hobbes* (Oxford: OUP, 2002).

Malkasian, C., *The American War in Afghanistan: A History* (Oxford: OUP, 2021).

Manningham-Buller, E., *Securing Freedom* (London: Profile Books, 2012).

Martin, M. J., and Sassner, C. W., *Predator: The Remote-Control Air War over Iraq and Afghanistan: A Pilot's Story* (Minneapolis: Zenith Press, 2010).

Matchett, W., *Secret Victory: The Intelligence War That Beat the IRA* (Lisburn: Hiskey Ltd, 2016).

Mayall, J., and Silvestri, S., *The Role of Religion in Conflict and Peace-Building* (London: British Academy, 2015).

McAllister, I., *The Northern Ireland Social Democratic and Labour Party: Political Opposition in a Divided Society* (London: Macmillan, 1977).

McAuley, J. W., *Very British Rebels? The Culture and Politics of Ulster Loyalism* (London: Bloomsbury, 2016).

McCall, C., *Border Ireland: From Partition to Brexit* (London: Routledge, 2021).

McCann, C., *Apeirogon: A Novel* (London: Bloomsbury, 2020).

McCleery, M. J., *Operation Demetrius and Its Aftermath: A New History of the Use of Internment without Trial in Northern Ireland, 1971–75* (Manchester: Manchester University Press, 2015).

McConaghy, K., *Terrorism and the State: Intra-State Dynamics and the Response to Non-State Political Violence* (Basingstoke: Palgrave Macmillan, 2017).

McConville, S., *Irish Political Prisoners 1848–1922: Theatres of War* (London: Routledge, 2003).

McConville, S., *Irish Political Prisoners, 1920–1962: Pilgrimage of Desolation* (London: Routledge, 2014).

Macdonald, A. [Pierce, W. L.], *The Turner Diaries* (Laurel Bloomery: The National Alliance, 1978).

McDonald, K., *Radicalization* (Cambridge: Polity Press, 2018).

McDonald, H., and Holland, J., *INLA: Deadly Divisions* (Dublin: Poolbeg Press, 1994).

McEvoy, K., *Paramilitary Imprisonment in Northern Ireland: Resistance, Management, and Release* (Oxford: OUP, 2001).

McGarry, F., *The Rising. Ireland: Easter 1916* (Oxford: OUP, 2010).

McGartland, M., *Fifty Dead Men Walking: The Heroic True Story of a British Secret Agent inside the IRA* (London: Blake 1998; 1st edn 1997).

McGladdery, G., *The Provisional IRA in England: The Bombing Campaign 1973–1997* (Dublin: Irish Academic Press, 2006).

McGlinchey, M., *Unfinished Business: The Politics of 'Dissident' Irish Republicanism* (Manchester: Manchester University Press, 2019).

McGovern, M., 'Informers, Agents, and the Liberal Ideology of Collusion in Northern Ireland', *Critical Studies on Terrorism*, 9/2 (2016).

McGuinness, M., *Bodenstown '86* (London: Wolfe Tone Society, n.d.).

McGuire, E., *Enemy of the Empire: Life as an International Undercover IRA Activist* (Dublin: O'Brien Press, 2006).

McIntyre, A., *Good Friday: The Death of Irish Republicanism* (New York: Ausubo Press, 2008).

McKearney, T., *The Provisional IRA: From Insurrection to Parliament* (London: Pluto Press, 2011).

McKittrick, D., Kelters, S., Feeney, B., and Thornton, C., *Lost Lives: The Stories of the Men, Women, and Children Who Died as a Result of the Northern Ireland Troubles* (Edinburgh: Mainstream Publishing, 1999).

McLoughlin, P. J., *John Hume and the Revision of Irish Nationalism* (Manchester: Manchester University Press, 2010).

McMaster, H. R., *Battlegrounds: The Fight to Defend the Free World* (London: William Collins, 2020).

Merari, A., *Driven to Death: Psychological and Social Aspects of Suicide Terrorism* (Oxford: OUP, 2010).

Merari, A., and Elad, S., *The International Dimension of Palestinian Terrorism* (Boulder: Westview Press, 1986).

Metzger, B. M., and Coogan, M. D. (eds), *The Oxford Companion to the Bible* (Oxford: OUP, 1993).

Milton-Edwards, B., and Farrell, S., *Hamas* (Cambridge: Polity Press, 2010).

Mishal, S., and Sela, A., *The Palestinian Hamas: Vision, Violence, and Coexistence* (New York: Columbia University Press, 2006; 1st edn 2000).

Mitchell, G. J., *Making Peace* (London: William Heinemann, 1999).

Mobley, B. W., *Terrorism and Counter-Intelligence: How Terrorist Groups Elude Detection* (New York: Columbia University Press, 2012).

Moloney, E., *A Secret History of the IRA* (London: Penguin, 2002).

Moloney, E., *Voices from the Grave: Two Men's War in Ireland* (London: Faber and Faber, 2010).

Montgomery, R., 'The Good Friday Agreement and a United Ireland', *Irish Studies in International Affairs*, 32/2 (2021).

Morag, N., 'Measuring Success in Coping with Terrorism: The Israeli Case', *Studies in Conflict and Terrorism*, 28/4 (2005).

Moran, J., 'Evaluating Special Branch and the Use of Informant Intelligence in Northern Ireland', *Intelligence and National Security*, 25/1 (2010).

Morris, B., *Righteous Victims: A History of the Zionist-Arab Conflict, 1881–1999* (New York: Alfred A. Knopf, 1999).

Morrison, D., *Then the Walls Came Down: A Prison Journal* (Cork: Mercier Press, 1999).

Morrison, J. F., *The Origins and Rise of Dissident Irish Republicanism: The Role and Impact of Organizational Splits* (New York: Bloomsbury, 2013).

Mueller, J., and Stewart, M. G., *Terror, Security, and Money: Balancing the Risks, Benefits, and Costs of Homeland Security* (Oxford: OUP, 2011).

Mueller, J., and Stewart, M. G., *Chasing Ghosts: The Policing of Terrorism* (Oxford: OUP, 2016).

Mulvenna, G., *Tartan Gangs and Paramilitaries: The Loyalist Backlash* (Liverpool: Liverpool University Press, 2016).

Mumford, A., *The West's War against Islamic State: Operation Inherent Resolve in Syria and Iraq* (London: I. B. Tauris, 2021).

Muro, D. (ed.), *When Does Terrorism Work?* (London: Routledge, 2018).

Muro, D., and Wilson, T. (eds), *Contemporary Terrorism Studies* (Oxford: OUP, 2022).

Murray, G., and Tonge, J., *Sinn Fein and the SDLP: From Alienation to Participation* (Dublin: O'Brien Press, 2005).

Nelson, S., *Ulster's Uncertain Defenders: Loyalists and the Northern Ireland Conflict* (Belfast: Appletree Press, 1984).

Netanyahu, B., *Fighting Terrorism: How Democracies Can Defeat Domestic and International Terrorists* (New York: Farrar Straus Giroux, 1995).

Neumann, P. R., *Radicalized: New Jihadists and the Threat to the West* (London: I. B. Tauris, 2016).

Neumann, P. R., *Bluster: Donald Trump's War on Terror* (London: Hurst and Co., 2019).

Neumann, P. R., and Smith, M. L. R., *The Strategy of Terrorism: How It Works, and Why It Fails* (London: Routledge, 2008).

Nixon, J., *Debriefing the President: The Interrogation of Saddam Hussein* (London: Bantam Press, 2016).

Norman, J. M., *The Second Palestinian Intifada: Civil Resistance* (London: Routledge, 2010).

Norton, A. R., *Hezbollah: A Short History* (Princeton: Princeton University Press, 2007).

O'Brien, J., *Killing Finucane: Murder in Defence of the Realm* (Dublin: Gill and Macmillan, 2005).

O'Callaghan, S., *The Informer* (London: Bantam, 1998).

Ó Dochartaigh, N., *Deniable Contact: Back-Channel Negotiation in Northern Ireland* (Oxford: OUP, 2021).

O'Doherty, M., *The Trouble with Guns: Republican Strategy and the Provisional IRA* (Belfast: Blackstaff Press, 1998).

O'Leary, B., *A Treatise on Northern Ireland. Volume 1—Colonialism: The Shackles of the State and Hereditary Animosities. Volume 2—Control: The Second Protestant Ascendancy and the Irish State. Volume 3—Consociation and Confederation: From Antagonism to Accommodation?* (Oxford: OUP, 2019).

O'Malley, E., *On Another Man's Wound* (Dublin: Anvil Books, 1979; 1st edn 1936).

Omand, D., *Securing the State* (London: Hurst and Co., 2010).

Omand, D., *How Spies Think: Ten Lessons in Intelligence* (London: Viking, 2020).

Omand, D., and Phythian, M., *Principled Spying: The Ethics of Secret Intelligence* (Oxford: OUP, 2018).

O'Mara, S., *Why Torture Doesn't Work: The Neuroscience of Interrogation* (Cambridge, MA: Harvard University Press, 2015).

Oren, M. B., *Six Days of War: June 1967 and the Making of the Modern Middle East* (Oxford: OUP, 2002).

Orwell, G., *Nineteen Eighty-Four* (London: Penguin, 1954 edn).

Pape, R. A., and Chicago Project on Security and Threats, *Understanding American Domestic Terrorism: Mobilization Potential and Risk Factors of a New Threat Trajectory* (Chicago: University of Chicago, 2021).

Parker, J. D., *On the Waterfront* (Durham: Pentland Press, 2000).

Parker, T., *Avoiding the Terrorist Trap: Why Respect for Human Rights Is the Key to Defeating Terrorism* (London: World Scientific Publishing, 2019).

Patterson, H., *The Politics of Illusion: A Political History of the IRA* (London: Serif, 1997; 1st edn 1989).

Pedahzur, A., *The Israeli Secret Services and the Struggle against Terrorism* (New York: Columbia University Press, 2010; 1st edn 2009).

Peleg, I., and Waxman, D., *Israel's Palestinians: The Conflict Within* (Cambridge: CUP, 2011).

Perl, R., *Combating Terrorism: The Challenge of Measuring Effectiveness* (Washington, DC: Congressional Research Service Report for Congress, 2007).

Perliger, A., *American Zealots: Inside Right-Wing Domestic Terrorism* (New York: Columbia University Press, 2020).

Pinker, S., *The Better Angels of Our Nature: The Decline of Violence in History and Its Causes* (London: Penguin, 2011).

Porter, P., *Blunder: Britain's War in Iraq* (Oxford: OUP, 2018).

Powell, J., *Great Hatred, Little Room: Making Peace in Northern Ireland* (London: Bodley Head, 2008).

Powell, J., *Talking to Terrorists: How to End Armed Conflicts* (London: Bodley Head, 2014).

Price, B. C., *Targeting Top Terrorists: Understanding Leadership Removal in Counter-Terrorism Strategy* (New York: Columbia University Press, 2019).

Qassem, N., *Hizbullah: The Story from Within* (London: Saqi, 2010; 1st edn 2005).

Ralph, J., *America's War on Terror: The State of the 9/11 Exception from Bush to Obama* (Oxford: OUP, 2013).

Rapoport, D. C., 'The Capitol Attack and the 5th Terrorism Wave', *Terrorism and Political Violence*, 33/5 (2021).

Rapoport, D. C., *Waves of Global Terrorism: From 1879 to the Present* (New York: Columbia University Press, 2022).

Reed, R. J., 'Blood, Thunder and Rosettes: The Multiple Personalities of Paramilitary Loyalism between 1971 and 1988', *Irish Political Studies*, 26/1 (2011).

Reinisch, D., 'Teenagers and Young Adults in Dissident Irish Republicanism: A Case Study of Na Fianna Éireann in Dublin', *Critical Studies on Terrorism*, 13/4 (2020).

Reynolds, A., *My Autobiography* (London: Transworld, 2009).

Richards, A., Margolin, D., and Scremin, N. (eds), *Jihadist Terror: New Threats, New Responses* (London: I. B. Tauris, 2019).

Richardson, L., *What Terrorists Want: Understanding the Terrorist Threat* (London: John Murray, 2006).

Rimington, S., *Open Secret: The Autobiography of the Former Director-General of MI5* (London: Hutchinson, 2001).

Rimington, S., *At Risk* (London: Arrow Books, 2004).

Rimington, S., *Secret Asset* (London: Arrow Books, 2006).

Risen, J., *Pay Any Price: Greed, Power, and Endless War* (Boston: Houghton Mifflin Harcourt, 2014).

Roberts, A., 'The "War on Terror" in Historical Perspective', *Survival*, 47/2 (2005).

Robson, L., *States of Separation: Transfer, Partition, and the Making of the Modern Middle East* (Oakland: University of California Press, 2017).

Roth, P., *Operation Shylock: A Confession* (London: Vintage, 2000; 1st edn 1993).

Rowan, B., *Political Purgatory: The Battle to Save Stormont and the Play for a New Ireland* (Newbridge: Merrion Press, 2021).

Rushdie, S., *Joseph Anton: A Memoir* (London: Jonathan Cape, 2012).

Ryder, C., *Inside the Maze: The Untold Story of the Northern Ireland Prison Service* (London: Methuen, 2000).

Ryder, C., *The RUC 1922–2000: A Force Under Fire* (London: Arrow Books, 2000).

Sacks, J., *Not in God's Name: Confronting Religious Violence* (London: Hodder and Stoughton, 2015).

Sanders, A., *Inside the IRA: Dissident Republicans and the War for Legitimacy* (Edinburgh: Edinburgh University Press, 2011).

Sanders, A., 'Northern Ireland: The Intelligence War 1969–1975', *British Journal of Politics and International Relations*, 3/2 (2011).

Sanders, A., *The Long Peace Process: The United States of America and Northern Ireland, 1960–2008* (Liverpool: Liverpool University Press, 2019).

Sanders, A., and Wood, I. S., *Times of Troubles: Britain's War in Northern Ireland* (Edinburgh: Edinburgh University Press, 2012).

Scheipers, S., *Unlawful Combatants: A Genealogy of the Irregular Fighter* (Oxford: OUP, 2015).

Schiemann, J. W., *Does Torture Work?* (Oxford: OUP, 2016).

Shafir, G., *A Half Century of Occupation: Israel, Palestine, and the World's Most Intractable Conflict* (Oakland: University of California Press, 2017).

Shapiro, J. N., *The Terrorist's Dilemma: Managing Violent Covert Organizations* (Princeton: Princeton University Press, 2013).

Shirlow, P., Tonge, J., McAuley, J., and McGlynn, C., *Abandoning Historical Conflict? Former Political Prisoners and Reconciliation in Northern Ireland* (Manchester: Manchester University Press, 2010).

Silber, M. D., *The Al-Qaida Factor: Plots against the West* (Philadelphia: University of Pennsylvania Press, 2012).

Silke, A. (ed.), *Routledge Handbook of Terrorism and Counter-Terrorism* (London: Routledge, 2019).

Sinai, J., Fuller, J., and Seal, T., 'Research Note: Effectiveness in Counter-Terrorism and Countering Violent Extremism: A Literature Review', *Perspectives on Terrorism*, 13/6 (2019).

Singh, R., *Hamas and Suicide Terrorism: Multi-Causal and Multi-Level Approaches* (London: Routledge, 2011).

Sky, E., *The Unravelling: High Hopes and Missed Opportunities in Iraq* (London: Atlantic Books, 2015).

M. Smith, *UDR Declassified* (Newbridge: Merrion Press, 2022).

Smith, M. L. R., and Jones, D. M., *The Political Impossibility of Modern Counter-insurgency: Strategic Problems, Puzzles, and Paradoxes* (New York: Columbia University Press, 2015).

Soufan, A. H., and Freedman, D., *The Black Banners (Declassified): How Torture Derailed the War on Terror after 9/11* (London: Penguin, 2020; 1st edn 2011).

Spencer, G. (ed.), *The British and Peace in Northern Ireland* (Cambridge: CUP, 2015).

Stalker, J., *Stalker* (London: Harrap, 1988).

Staniforth, A., and Sampson, F. (eds), *The Routledge Companion to UK Counter-Terrorism* (London: Routledge, 2013).

Stritzke, W. G. K., Lewandowsky, S., Denemark, D., Clare, J., and Morgan, F. (eds), *Terrorism and Torture: An Interdisciplinary Perspective* (Cambridge: CUP, 2009).

Tankel, S., *With Us and Against Us: How America's Partners Help and Hinder the War on Terror* (New York: Columbia University Press, 2018).

Taylor, I., *The Ethics of Counter-Terrorism* (London: Routledge, 2018).

Tembo, E. B., *US-UK Counter-Terrorism after 9/11: A Qualitative Approach* (London: Routledge, 2014).

Townshend, C., *Making the Peace: Public Order and Public Security in Modern Britain* (Oxford: OUP, 1993).

Townshend, C., *When God Made Hell: The British Invasion of Mesopotamia and the Creation of Iraq, 1914–1921* (London: Faber and Faber, 2010).

Townshend, C., *The Republic: The Fight for Irish Independence* (London: Penguin, 2013).

Townshend, C., *Terrorism: A Very Short Introduction* (Oxford: OUP, 2018; 1st edn 2002).

Townshend, C., *The Partition: Ireland Divided, 1885–1925* (London: Penguin, 2021).

Treverton, G. F., *Intelligence for an Age of Terror* (Cambridge: CUP, 2009).

Tripp, C., *A History of Iraq* (Cambridge: CUP, 2007).

Tuck, R., *Hobbes* (Oxford: OUP, 1989).

Tuck, R., *The Rights of War and Peace: Political Thought and the International Order from Grotius to Kant* (Oxford: OUP, 1999).

Tucker, D., and Lamb, C. J., *United States Special Operations Forces* (New York: Columbia University Press, 2007).

Veilleux-Lepage, Y., *How Terror Evolves: The Emergence and Spread of Terrorist Techniques* (London: Rowman and Littlefield, 2020).

von Clausewitz, C., *On War* (London: Penguin, 1968; 1st edn 1832).

Waldron, J., *Torture, Terror, and Trade-Offs: Philosophy for the White House* (Oxford: OUP, 2010).

Waller, J., *A Troubled Sleep: Risk and Resilience in Contemporary Northern Ireland* (Oxford: OUP, 2021).

Walter, B. F., *How Civil Wars Start: And How to Stop Them* (London: Viking, 2022).

Whelehan, N., *The Dynamiters: Irish Nationalism and Political Violence in the Wider World, 1867–1900* (Cambridge: CUP, 2012).

White, R. W., *Ruairí Ó Brádaigh: The Life and Politics of an Irish Revolutionary* (Bloomington: Indiana University Press, 2006).

White, R. W., *Out of the Ashes: An Oral History of the Provisional Irish Republican Movement* (Newbridge: Merrion Press, 2017).

White, R. W., Demirel-Pegg, T., and Lulla, V., 'Terrorism, Counter-Terrorism, and the "Rule of Law": State Repression and "Shoot-to-Kill" in Northern Ireland', *Irish Political Studies*, 36/2 (2021).

Whiting, S. A., *Spoiling the Peace? The Threat of Dissident Republicans to Peace in Northern Ireland* (Manchester: Manchester University Press, 2015).

Wight, C., *Rethinking Terrorism: Terrorism, Violence, and the State* (London: Palgrave Macmillan, 2015).

Wilkinson, P., *Terrorism versus Democracy: The Liberal State Response* (London: Routledge, 2011; 1st edn 2001).

Williams, B. G., *Counter Jihad: America's Military Experience in Afghanistan, Iraq, and Syria* (Philadelphia: University of Pennsylvania Press, 2017).

Williams, E. L., and Leahy, T., 'The "Unforgiveable"?: Irish Republican Army (IRA) Informers and Dealing with Northern Ireland Conflict Legacy, 1969–2021', *Intelligence and National Security*, published online 3 August 2022, https://www-tandfonline-com.queens.ezp1.qub.ac.uk/doi/full/10.1080/02684527.2022.2104000.

Wills, D. C., *The First War on Terrorism: Counter-Terrorism Policy during the Reagan Administration* (Lanham: Rowman and Littlefield, 2003).

Wilson, G., *Marie: A Story from Enniskillen* (London: Marshall Pickering, 1990).

Wilson, R., and Adams, I., *Special Branch—A History: 1883–2006* (London: Biteback, 2015).

Wilson, T. K., *Killing Strangers: How Political Violence Became Modern* (Oxford: OUP, 2020).

Wood, G., *The Way of the Strangers: Encounters with the Islamic State* (London: Penguin, 2018; 1st edn 2017).

Wood, I. S., *Crimes of Loyalty: A History of the UDA* (Edinburgh: Edinburgh University Press, 2006).

Woodford, I., and Smith, M. L. R., 'The Political Economy of the Provos: Inside the Finances of the Provisional IRA—A Revision', *Studies in Conflict and Terrorism*, 41/3 (2018).

Wright, L., *The Looming Tower: Al-Qaida's Road to 9/11* (London: Penguin, 2007; 1st edn 2006).

Yousef, M. H. and Brackin, R., *Son of Hamas* (Milton Keynes: Authentic Media Limited, 2014; 1st edn 1999).

Zegart, A. B., *Flawed by Design: The Evolution of the CIA, JCS, and NSC* (Stanford: Stanford University Press, 1999).

Zegart, A. B., *Spying Blind: The CIA, the FBI, and the Origins of 9/11* (Princeton: Princeton University Press, 2007).

Zellen, B. S., *State of Recovery: The Quest to Restore American Security after 9/11* (London: Bloomsbury, 2013).

Zoller, S., *To Deter and Punish: Global Collaboration against Terrorism in the 1970s* (New York: Columbia University Press, 2021).

Zulaika, J., *Terrorism: The Self-Fulfilling Prophecy* (Chicago: University of Chicago Press, 2009).

Zulaika, J., *Hellfire from Paradise Ranch: On the Frontlines of Drone Warfare* (Oakland: University of California Press, 2020).

Index